Stochastic Opinion Dynamics: Theory and Application

苏 伟　陈 鸽　著

Beijing
Metallurgical Industry Press
2021

Abstract

Stochastic drive plays a key role in generating synchronization behaviors of bounded confidence opinion dynamics. This book presents the new results on the noise-induced synchronization of bounded confidence opinion dynamics. It consists of 8 chapters, including both the basic theories of noise-induced synchronization of opinion dynamics and the applications of the theories to important sociological problems. This results meanwhile help reveal the fundamental mechanism of noise-driven order in self-organizing complex systems.

This book is written for researchers and engineers working in systems theory and control, complex systems, social networks, and machine learning etc.

图书在版编目(CIP)数据

随机舆论动力学：理论及应用 = Stochastic Opinion Dynamics：Theory and Application：英文/苏伟，陈鸽著. —北京：冶金工业出版社，2021.3

ISBN 978-7-5024-8754-6

Ⅰ.①随… Ⅱ.①苏… ②陈… Ⅲ.①随机噪声—动力学—研究—英文 Ⅳ.①O211.6

中国版本图书馆 CIP 数据核字(2021)第 044530 号

出 版 人 苏长永
地　　址 北京市东城区嵩祝院北巷 39 号 邮编 100009 电话 (010)64027926
网　　址 www.cnmip.com.cn 电子信箱 yjcbs@cnmip.com.cn
责任编辑 戈 兰 美术编辑 彭子赫 版式设计 孙跃红
责任校对 石 静 责任印制 李玉山
ISBN 978-7-5024-8754-6
冶金工业出版社出版发行；各地新华书店经销；北京虎彩文化传播有限公司印刷
2021 年 3 月第 1 版，2021 年 3 月第 1 次印刷
169mm×239mm；8.25 印张；157 千字；121 页
56.00 元

冶金工业出版社 投稿电话 (010)64027932 投稿信箱 tougao@cnmip.com.cn
冶金工业出版社营销中心 电话 (010)64044283 传真 (010)64027893
冶金工业出版社天猫旗舰店 yjgycbs.tmall.com
（本书如有印装质量问题，本社营销中心负责退换）

Preface

Understanding the origin and the underlying principles of synchronization phenomena in self-organizing systems has been a grand transdisciplinary challenge. A particularly vibrant and recently emerging novel line of research in this area has been dedicated to the study of noise-induced synchronization in human collective behavior such as global opinion formation. However, due to the inherent difficulty of its mathematical analysis, most theoretical approaches largely neglected the influence of noise on synchronization processes. Recently, the theoretical development of noise-driven synchronization of bounded confidence opinion dynamics provides a window for people to look into how stochastic effect induces the order of a system from a disordered state.

This book presents the new results on the noise-induced synchronization of bounded confidence opinion dynamics. Bounded confidence implies that an agent updates its states by only considering of the states within its confidence threshold. Hegselmann-Krause (HK) model, a standard model of bounded confidence, is taken to display the methodologies and the main theoretical development of the topic. Also, some applications of positive noise effect in social meaning are collected for the readers to understand this topic. The book is written for researchers and engineers working in systems theory and control, complex systems, social networks, and machine learning etc.

This book could not have been completed without the help and encouragement of many people. We recognize our institutions and colleagues for providing us with a stimulating and supportive academic environment. We are deeply indebted to many researchers in the field for insightful discussions and constructive criticisms, and for enriching us with their expertise and enthusiasm. This book is supported by the National Natural Science Foundation of China under Grants 11688101, 12071465, 61803024, the National Key Basic Research Program of China (973 program) under grant 2016YFB0800404 and the National Key Research and Development Program of Ministry of Science and Technology of China under Grant 2018AAA0101002.

Wei Su, Ge Chen
Beijing, October 2020

Contents

1 Introduction

1.1 Motivation and Background

Opinion dynamics is a research field in which various tools are used to study the dynamical processes of the formation, diffusion, and evolution of public opinion. In fact, opinion dynamics has been an important issue of research in sociology and has also attracted a lot of attention in recent years from many other disciplines such as physics, mathematics, computer science, social psychology, and philosophy[1,2].

The study of opinion dynamics can be traced back to the two-step flow of communication model studied by Lazarsfeld and Katz in the 1940-50s[3]. This model posits that most people form their opinions under the influence of opinion leaders, who, in turn, are influenced by the mass media. Another famous early work on opinion dynamics is the social power model proposed by French[4]. Based on a discussion and classification of "social power", this model describes the diffusion of social influence and the formation of public opinions in social networks. The French model is a special case of the model proposed by DeGroot[5]; as such, they are referred to as the "French-DeGroot model" in some works[6]. Later, some new theories of opinion dynamics have been developed, namely the social influence network theory[7], social impact theory[8], and dynamic social impact theory[9]. Recently, bounded confidence (BC) models of opinion dynamics has been of interest. The BC models adopt a mechanism where one individual is not willing to accept the opinion of another one if he/she feels their opinions have a big gap. One well-known BC model was formulated by Hegselmann and Krause[10], called the Hegselmann-Krause (HK) model, where all agents synchronously update their opinions by averaging the opinions in their confidence bounds. Another well-known BC model, called Deffuant-Weisbuch (DW) model, was proposed by Deffuant and Weisbuch[11], which is similar to the HK model, though it instead employs a pairwise-sequential updating procedure in place of the synchronized one within the HK model.

Generally, the above models can be divided into two classes. Except the BC models, mainly including the HK model and the DW model, the "French-DeGroot model" as well as the Friedkin-Johnsen[12] model, can be interpreted as the topology-based models. The BC mechanism assumes that each individual in the system has a bounded confidence and only takes into account those opinions for updating that are located within a given confidence level; on the other hand, the topology-based mechanism preassigns the interaction relationship among individuals, i. e. their opinions (irrespective of whether it is time-varying or time-invariant). In the original opinion models, once the initial opinion values are given, the opinions are assumed to evolve in a deterministic way. At the same time, the theoretical studies mainly focus on the deterministic case of these models.

Stochastic opinion dynamics possesses its significance in two aspects. For one thing, the actual opinions of individuals may fluctuate stochastically during evolution due to personal free will or the random influence coming from mass media. Moreover, as Sagués et al. explained in[13], "natural systems are undeniably subject to random fluctuations, arising from either environmental variability or thermal effects". Hence, it is a natural question that how opinion evolves in a noisy environment. For another, some profound reasons may involve. For the noisy topology-based models, by the robust consensus theory of noisy multi-agent systems it is ready to know that the group can achieve robust consensus if and only if a *jointly connected* condition is satisfied[14, 15]. When turning to the BC models in noisy environment, some interesting phenomena were discovered in the simulation studies[16~20]. Instead of the damage effect which noise plays in most circumstances, it is found that noise plays a positive role in enhancing the consensus of opinions with BC dynamics. The second fact largely coincides with the principle of *noise induced order*, which exists ubiquitously in the synchronization of self-organizing systems.

In the past decades, self-organizing systems based on local rules have been used to investigate the collective behavior in natural and social systems, and several models have been proposed, including the widely known Boid and Vicsek models[21, 22]. Understanding the origin and the underlying principles of synchronization phenomena in self-organizing systems has been a grand transdisciplinary challenge[23]. The inclination of natural and artificial systems to synchronize such that their component units become locked in time may represent one of the most pervasive urges in the fabric of the universe[24, 25]. After Heinz von Foerster proposed the principle of "order from noise" in 1960[26], noise has been believed to be a key factor in promoting the synchronization of self-organizing systems, which has been verified in earlier simulation or experimental studies in a number of fields[13, 27~33].

To essentially reveal the synchronization mechanism of self-organizing systems and explore the "noise-induced order" principle, it is necessary to theoretically analyze the noisy dynamics of self-organizing systems. Due to the difficulty of analysis, most previous theoretical studies on synchronization of self-organizing systems largely ignored the influence of noise[34~37]. Only lately, some mathematical analysis appears. For example, the analysis of the Vicsek model subject to noise was first carried out by Chen in 2017[38]. Beyond that, a substantial and systematic mathematical study on how noise affects the synchronization of self-organizing systems has been infrequent. Until recently, the analysis of noisy HK opinion dynamics starts a new journey towards the mathematical study of noise driven order.

For the noisy BC opinion dynamics, especially the HK model, several simulations has been conducted[16~20]; however, few strictly theoretical results have been obtained ever since. The main difficulty in the analysis of the noisy HK model lies in that the inter-agent topology is time-varying and determined by the agents' states, whereas the agents' states are dependent on topology and noise. Huang et al proposed a HK-based stochastic approximation model with damped noises and analyzed the properties of stable points[39]. Wang and Chazelle studied a continuous noisy HK model. In[40], Su et al gave a complete analysis of the noise-induced consensus based on the original HK model. It is proved that the HK opinion dynamics can achieve a *quasi-consensus* from any initial state under the drive of noise, even through the system cannot achieve consensus without noise. Moreover, a critical noise strength is obtained. After that, Su et al obtained a series of results within the range of noise-induced consensus of HK-like models.

This book presents the main results currently obtained in the analysis of stochastic opinion dynamics generated by HK-like models in noisy environment. For the noisy topology-based models, the analysis is similar to the traditional multi-agent systems in noisy environment. The results collected in this book include the basic theoretical conclusions of noisy HK models. These results establish the foundation of the theory of noise-induced synchronization of BC dynamics. Other than that, the theory is applied to analyze the issues in social opinions, such as truth seeking, generation of social cleavage, and social opinion intervention etc.

1.2 Outline of the Book

This book can be roughly divided into two parts: the first part is about the theory of noise-induced synchronization of HK-like models (chapters 3 ~ 5), and the second part shows the application of the theory (chapters 6 ~ 8).

Chapter 2 prepares the mathematical tools used in the book, mainly including the

Bore-Cantelli theorem, limit properties of the sum of independent random variables, and theory of stopping time.

Chapter 3 introduces the basic homogeneous HK model with noise. To keep the same with the original HK model, the noisy model in this chapter is also limited to a bounded state space. Then by defining the consensus in noisy case, namely *quasi-consensus*, the basic theoretical analysis of the noise-induced synchronization is given. It proves that for any initial state, the system can almost surely achieve quasi-consensus in a finite time under the drive of noise. Moreover, a critical noise strength is obtained. When the noise strength is no larger than the critical value, the noisy system can always achieve quasi-consensus; when the noise strength exceeds the critical value, the systems can never achieve quasi-consensus.

Chapter 4 extends the noisy homogeneous HK model in bounded space to the one in unbounded space. The modified model allows the system state to move in the full space and hence largely enhance the applicability of HK dynamics in modelling of self-organizing systems. However, this modification results in tougher mathematical difficulties in proving the conclusion. Using the techniques of independent stopping times, this chapter finally obtains the similar results as those in Chapter 3. Moreover, a much more elegant critical noise strength is given in this chapter.

Chapter 5 establishes the theory of heterogeneous HK dynamics with both environment and communication noise. The heterogeneous model is more important than the homogeneous one, since it allows agents to possess heterogeneous personal inclination. However, the analysis of the heterogeneous HK model is much more difficult. The convergence of the general heterogeneous HK model is still an open problem except only some partial results. The chapter provides the main theoretical results on the noisy heterogeneous HK model.

Chapter 6 starts to present the applications of the theory of noise-induced synchronization to social issues. This chapter first uses a BC-based truth-seeking model to show that noise can make all the agents, including truth seekers and blind followers, find truth. Here, noise can model the free information flow in a society. The conclusion in this chapter verifies a sociological discovery, namely that free information flow is beneficial to gaining truth.

Chapter 7 designs a noise-based intervention strategy to eliminate social disagreement. The noise intervention is applied to only one agent, and it is proved that the existing disagreement in a group is eliminated in a finite time. Moreover, after introducing studying the first passage time of a new random walk, the finite stopping time when the group achieves agreement is precisely calculated.

Chapter 8 investigates how social cleavage is generated. Previously, bounded confidence was considered as a key factor that produces opinion fragmentation. However, the results in former chapters reveal that in the presence of noise, the fragmentation of bounded confidence models disappears finally. In this chapter, we find that it is the inherent heterogeneous prejudices within society, instead of mere bounded confidence, that bring social cleavage in an information age.

References

[1] Lorenz J. Continuous opinion dynamics under bounded confidence: A survey[J]. International Journal of Modern Physics C, 2007, 18 (12): 1819 ~ 1838.

[2] Sîrbu A, Loreto V, Servedio V D, Tria F. Opinion dynamics: models, extensions and external effects[M]. In Participatory Sensing, Opinions and Collective Awareness (pp. 363 ~ 401). Springer International Publishing, 2017.

[3] Katz E, Lazarsfeld F. Personal Influence: the Part Played by People in the Flow of Mass Communications[M]. Free Press, Glencoe, IL, 1955.

[4] French J R. A formal theory of social power[J]. Psychological review, 1956, 63 (3): 181 ~ 194.

[5] DeGroot M H. Reaching a consensus[J]. Journal of American Statistical Association, 1974, 69 (345):118 ~ 121.

[6] Proskurnikov A V, Tempo R. A tutorial on modeling and analysis of dynamic social networks Part I [J]. Annual Reviews in Control, 2017, 43: 65 ~ 79.

[7] Friedkin N. A structural theory of social influence[J]. Cambridge University Press, Cambridge, UK, 1998.

[8] Latané B. The psychology of social impact, American Psychologist[J] 1981, 36: 343 ~ 365.

[9] Latané B. Dynamic social impact: The creation of culture by communication[J]. Journal of Communication, 1996, 46 (4): 13 ~ 25.

[10] Hegselmann R, Krause U. Opinion dynamics and bounded confidence models, analysis, and simulation[J]. Artificial Societies and Social Simulation, 2002, 5 (3): 1 ~ 33.

[11] Deffuant G, Neau D, Amblard F, Weisbuch G. Mixing beliefs among interacting agents[J]. Advances in Complex Systems, 2000, 3 (01n04): 87 ~ 98.

[12] Friedkin N, Johnsen E. Social influence networks and opinion change[J]. Adv. Group Proc, 1999, 16 (1): 1 ~ 29.

[13] Sagués F, Sancho J M, Garca-Ojalvo J. Spatiotemporal order out of noise[J]. Reviews of Modern Physics, 2007, 79 (3): 829 ~ 882.

[14] Wang L, Liu Z, Guo L. Robust consensus of multi-agent systems with noise[C]. Proceedings of the 26th Chinese Control Conference, CCC 2007, 2007, 737 ~ 741.

[15] Wang L, Guo L. Robust consensus and soft control of multi-agent systems with noises[J]. Jour-

nal of Systems Science and Complexity, 2008, 21 (3): 406 ~ 415.

[16] Mäs M, Flache A, Helbing D. Individualization as driving force of clustering phenomena in humans[J]. PLoS Computational Biology, 2010, 6 (10), e1000959.

[17] Pineda M, Toral R, Hernndez-Garca E. Diffusing opinions in bounded confidence processes[J]. The European Physical Journal D, 2011, 62 (1): 109 ~ 117.

[18] Grauwin S, Jensen P. Opinion group formation and dynamics: Structures that last from nonlasting entities[J], Physical Review E, 2012, 85 (6), 066113.

[19] Carro A, Toral R, Miguel M S. The role of noise and initial conditions in the asymptotic solution of a bounded confidence, continuous-opinion model[J]. Journal of Statistical Physics, 151 (1 - 2): 131 ~ 149, 2013.

[20] Pineda M, Toral R, Hernndez-Garca E. The noisy Hegselmann-Krause model for opinion dynamics[J]. The European Physical Journal B, 2013, 86 (12): 1 ~ 10.

[21] Reynolds C W. Flocks, herds and schools: A distributed behavioral model[J], ACM Siggraph Computer Graphics,1987, 21 (4): 25 ~ 34.

[22] Vicsek T, Czirk A, Ben-Jacob E, Cohen I, Shochet O. Novel type of phase transition in a system of self-driven particles[J]. Physical Review Letters, 1995, 75 (6): 1226 ~ 1229.

[23] Strogatz S. Sync: How order emerges from chaos in the universe, nature, and daily life Hachette Books[J]. New York, 2003.

[24] Collins J J. Move to the rhythm[J]. Nature, 2003, 422: 117 ~ 118.

[25] N'eda Z, Ravasz E, Brechet Y, Vicsek T, Barabási A L. The sound of many hands clapping [J]. Nature, 2000, 403: 849 ~ 850.

[26] Von Foerster H. On self-organizing systems and their environments 31-50 in Self-organizing systems[M]. M. C. Yovits and S. Cameron (eds.), Pergamon Press, London, 1960.

[27] Shinbrot T, Muzzio F J. Noise to order[J]. Nature, 2001, 410 (6825): 251 ~ 258.

[28] Matsumoto K, Tsuda I, Noise-induced order[J]. Journal of Statistical Physics, 1983, 31 (1): 87 ~ 106.

[29] Eldar A, Elowitz M B. Functional roles for noise in genetic circuits[J]. Nature, 2010, 467 (7312): 167 ~ 173.

[30] Tsimring L S. Noise in biology, Reports on progress in physics Physical Society (Great Britain), 2014, 77 (2), 026601.

[31] Zhou T, Chen L, Aihara K. Molecular communication through stochastic synchronization induced by extracellular fluctuations[J]. Physical Review Letters, 2005, 95 (17), 178103.

[32] Lichtenegger K, Hadzibeganovic T. The interplay of self-reflection, social interaction and random events in the dynamics of opinion flow in two-party democracies[J]. International Journal of Modern Physics C, 2016, 27, 1650065.

[33] Shirado H, Christakis N A. Locally noisy autonomous agents improve global human coordination in network experiments[J]. Nature, 2017, 545, 370374.

[34] Jadbabaie A, Lin J, Morse A S. Coordination of groups of mobile autonomous agents using nearest neighbor rules[J]. IEEE Transactions on Automatic Control, 2003, 48 (6): 988 ~ 1001.

[35] Savkin A V. Coordinated collective motion of groups of autonomous mobile robots: Analysis of Vicsek's model[J]. IEEE Transactions on Automatic Control, 2004, 49 (6): 981 ~983.

[36] Tang G, Guo L. Convergence of a class of multi-agent systems in probabilistic framework[J]. Journal of Systems Science and Complexity, 2007, 20 (2): 173 ~ 197.

[37] Chen G, Liu Z, Guo L. The smallest possible interaction radius for synchronization of self-propelled particles[J]. SIAM Review, 2014, 56 (3): 499 ~521.

[38] Chen G. Small noise may diversify collective motion in Vicsek model[J]. IEEE Transactions on Automatic Control, 2017, 62 (2): 636 ~651.

[39] Huang M, Manton J. Opinion dynamics with noisy information[C]. Proceedings of the IEEE Conference on Decision and Control, 2003, 3445 ~3450.

[40] Su W, Chen G, Hong Y. Noise leads to quasi-consensus of Hegselmann-Krause opinion dynamics[J]. Automatica, 2017, 85: 448 ~454.

2 Preliminaries

In this chapter, some results of probability theory are introduced as the tools of analysis in this book.

2.1 Basic Definitions

σ-algebra Let Ω be a space, $\mathcal{F}$ is a nonempty class of subsets of Ω. $\mathcal{F}$ is called a σ-algebra if

(1) $A^c \in \mathcal{F}$ whenever $A \in \mathcal{F}$,

(2) $\bigcup_{n=1}^{\infty} A_n \in \mathcal{F}$ whenever $A_n \in \mathcal{F}, n \geqslant 1$.

Probability Space Let $\mathcal{F}$ be a σ-algebra of subsets of a sample space Ω, and $\mathbb{P}$ is a measure on $\mathcal{F}$. If $\mathbb{P}\{\Omega\} = 1$, $\mathbb{P}$ is called a probability measure, and the measure space $(\Omega, \mathcal{F}, \mathbb{P})$ is a probability space. The sets $A \in \mathcal{F}$ are called events and the non-negative, real number $\mathbb{P}\{A\}$ is referred to as the probability of the event A, An event A is said to occur almost surely (abbreviated a. s.), if $\mathbb{P}\{A\} = 1$.

Random Variable A real-valued measurable function X on a probability space $(\Omega, \mathcal{F}, \mathbb{P})$ is called a random variable (abbreviated r. v.) if $\mathbb{P}\{|X| < \infty\} = 1$.

Independence If $(\Omega, \mathcal{F}, \mathbb{P})$ is a probability space and T a nonempty index set, classes $\{\mathcal{G}_t\}$ of events, $t \in T$, are termed independent if for each $m = 2, 3, \cdots$, each choice of distinct $t_j \in T$, and events $A_j \in \mathcal{G}_{t_j}$, $1 \leqslant j \leqslant m$

$$\mathbb{P}\left(\bigcap_{j=1}^{m} A_j\right) = \prod_{j=1}^{m} \mathbb{P}\{A_j\}$$

Events $\{A_t,\ t \in T\}$, are called independent if the one-element classes $\{\mathcal{G}_t = \{A_t\},\ t \in T\}$, are independent.

$\{X_n,\ n \geqslant 1\}$ are termed independent random variables if the classes $\{\mathcal{F}_n = \sigma(X_n),\ n \geqslant 1\}$ are independent. Here, $\mathcal{F}_n = \sigma(X_n)$, is the σ-algebra generated by X_n, $n \geqslant 1$.

Tail σ-algebra The tail σ-algebra of a sequence $\{X_n,\ n \geqslant 1\}$ of r. v. s on a probability space $(\Omega, \mathcal{F}, \mathbb{P})$ is $\bigcap_{n=1}^{\infty} \sigma(X_j,\ j \geqslant n)$. The sets of the tail σ-algebra are called tail events and functions measurable relative to the tail σ-algebra are called tail functions.

Stopping Time Let $(\Omega, \mathcal{F}, \mathbb{P})$ is a probability space and $\{\mathcal{F}_n, n \geqslant 1\}$ an increasing sequence of sub-σ-algebras of $\mathcal{F}$, that is, $\mathcal{F}_1 \subset \mathcal{F}_2 \subset \cdots \subset \mathcal{F}_n \subset \mathcal{F}$. A measurable function $T = T(\omega)$ taking values $1, 2, \cdots, \infty$ is called a stopping time relative to $\{\mathcal{F}_n\}$ or simply an $\{\mathcal{F}_n\}$ - time if $\{T = j\} \in \mathcal{F}_j$, $j = 1, 2\cdots$

2.2 Main Theorems

The following Borel-Cantelli theorem is elemental in probability of theory.

Theorem 2.1 (Borel-Cantelli) If $\{A_n, n \geqslant 1\}$ is a sequence of events with $\sum_{n=1}^{\infty} A_n < \infty$, then $\mathbb{P}\{A_n, i.o.\} = 0$. Conversely, if the events $\{A_n, n \geqslant 1\}$ are independent and $\sum_{n=1}^{\infty} A_n = \infty$, then $\mathbb{P}\{A_n, i.o.\} = 1$. Here, $\{A_n, i.o.\} = \bigcap_{n=1}^{\infty} \bigcup_{j=n}^{\infty} A_j$.

In studying the collective behavior of multi-agent systems with noise, it usually requires to deal with the sum of independent r. v. s. The following theorems are the classical results about independent r. v. s. [1].

Theorem 2.2 (Strong Law of Large Numbers) Let $\{X_n, n \geqslant 1\}$ be independent r. v. s, $\mathbf{E}X_n = 0$ and there exists $\alpha_n \in [1, 2]$ satisfying

$$\sum_{n=1}^{\infty} \frac{\mathbf{E}|X_n|^{\alpha_n}}{n^{\alpha_n}} < \infty$$

then

$$\frac{1}{n} \sum_{j=1}^{n} X_j \xrightarrow[n \to \infty]{a.s.} 0$$

Theorem 2.3 (Law of the Iterated Logarithm) Let $\{X_n, n \geqslant 1\}$ be independent r. v. s with $\mathbf{E}X_n = 0$, $\mathbf{E}X_n^2 = \sigma_n^2, s_n^2 = \sum_{i=1}^{n} \sigma_i^2 \to \infty$. If $|X_n| \leqslant d_n$, a. s., where the constant $d_n = o\left(\frac{s_n}{(\log\log s_n)^{1/2}}\right)$, as $n \to \infty$, then, setting $S_n = \sum_{i=1}^{n} X_i$

$$\mathbb{P}\left\{\overline{\lim_{n \to \infty}} \frac{S_n}{s_n \sqrt{\log\log s_n}} = \sqrt{2}\right\} = 1 = \mathbb{P}\left\{\underline{\lim_{n \to \infty}} \frac{S_n}{s_n \sqrt{\log\log s_n}} = -\sqrt{2}\right\}$$

Theorem 2.4 (Central Limit Theorems) $\{X_n, n \geqslant 1\}$ are independent r. v. s, with

$\mathbf{E}X_n = 0$. Let $s_n^2 = \sum_{j=1}^{n} \mathbf{E}X_j^2$, if $\sum_{j=1}^{n} \mathbf{E}|X_j|^{2+\delta} = o(s_n^{2+\delta})$ for some $\delta > 0$

then

$$\lim_{n\to\infty} \mathbb{P}\left\{\frac{S_n}{s_n} < x\right\} = \frac{1}{\sqrt{2\pi}}\int_{-\infty}^{x} \mathrm{e}^{-u^2/2} du$$

References

[1] Chow Y, Teicher H. Probability Theory: Independence, Interchange-ability, Martingales[M]. Springer Science & Business Media, 1997.

3 Homogeneous HK Model in Bounded Space

In this chapter, we aim to provide a theoretical analysis to investigate the consensus behavior of bounded HK opinion dynamics in noisy environments. In the HK model, once the initial opinion values are given, the opinions evolve in a deterministic way[1,2]. However, the actual opinions of individuals may fluctuate during the transmission and evolution of opinions because of free will or the influence of the social media or personal events. Therefore, how noise affects the behavior of opinion dynamics is a natural problem; it has been studied using simulations[3~7], with most of them demonstrating that noise may play a positive role in enhancing the consensus or reducing the disagreement of opinions.

For the noisy HK model, several simulations have been conducted[7]; however, few strictly theoretical results have been obtained. The main difficulty in the analysis of the noisy HK model lies in that the inter-agent topology is time-varying and determined by the agents' states, whereas the agents' states are dependent on topology and noise. Huang and Manton proposed a HK-based stochastic approximation model with damped noises and analyzed the properties of stable points[8]; however, this type of system differs from our noisy HK model to a certain degree. In addition, many studies have been conducted on average-consensus protocols with switching topologies and additive noises[9~12]; however almost all of these studies need to add some connectivity assumptions on topologies, which are difficult to verify in the noisy HK model because its topology is determined by the agents' states.

To provide a formal description, this chapter first defines the quasi-consensus in the noisy case, and then present a "critical phenomenon". Namely, when the noise strength is no larger than half the confidence threshold, the noisy HK model with arbitrary initial states can almost surely achieve quasi-consensus in finite time; otherwise, even the clustered opinions can get divided. This result not only reveals the evolutionary mechanism of opinions, but also assists in intervention strategy design to induce the opinion agreement by injecting random noise to social groups. In addition, HK dynamics is known to explain the divergence of opinions. However, the results reveal that the fragmentation behavior of the HK model fails to exhibit robustness against arbitrary weak random noise and that the mechanism of opinion divergence requires further study.

3.1　HK Models

The original homogeneous HK model has the following evolution dynamics[1]:

$$x_i(t+1) = \frac{1}{|N(i,x(t))|} \sum_{j \in N(i,x(t))} x_j(t) \tag{3.1}$$

where $i \in V = \{1,2,\cdots,n\}$, $x_i(t) \in [0,1]$, is the opinion value of agent i at time t and

$$N(i,\ x(t)) = \{j \in V \mid |x_j(t) - x_i(t)| \leqslant \varepsilon\} \tag{3.2}$$

is the neighbor set of agent i at t with $\varepsilon \in (0,1]$ representing the confidence threshold (or interaction radius) and $|\cdot|$ denoting the absolute value of a real number or the cardinal number of a set accordingly. Although it seems simple, the HK model (3.1) captures a common evolution mechanism of many practical social systems and also presents rich phenomena of opinion dynamics. Sometimes, all opinions gather in one cluster and achieve consensus or agreement, but very often, opinions split into more than one cluster, called fragmentation, which yields the disagreement of opinions. A standard opinion evolution of the HK model can be found in Fig. 3.1, where the opinion fragmentation appears with 3 opinion subgroups.

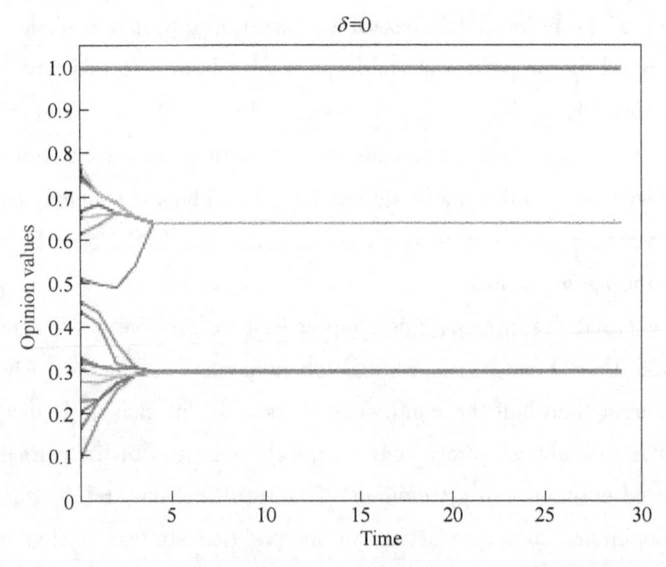

Fig. 3.1　Opinion evolution of system (3.1) with 20 agents

(The initial states are randomly generated from the interval [0,1] and the confidence threshold ε is chosen to be 0.2)

In reality, the evolution of an individual's opinion may be affected by the random or time-varying environment and unexpected accidents. Opinion evolution with random factors involves complexity; regardless, some studies have been conducted for simple cases in the literature (e.g. [3, 4, 8]). We also simplify the environmental disturbance to examine a modified HK model with persistent random noise. Opinion values are usually bounded because extreme ones are often finite. Without loss of generality, the 2 extreme opinion values in the noisy HK model are 0 and 1, and the other opinion values are limited between 0 and 1. The boundedness assumption is widely used in the study of noisy opinion dynamics[3, 7]. Following this assumption, the noisy HK model is described as follows:

$$x_i(t+1) = \begin{cases} 1, & x_i^*(t) > 1 \\ x_i^*(t), & x_i^*(t) \in [0,1], \ \forall i \in V, \ t \geqslant 0 \\ 0, & x_i^*(t) < 0 \end{cases} \tag{3.3}$$

where

$$x_i^*(t) = \frac{1}{|N(i, x(t))|} \sum_{j \in N(i,x(t))} x_j(t) + \xi_i(t+1) \tag{3.4}$$

with $\{\xi_i(t)\}_{i \in V, t>0}$ being the environmental noise.

3.2 Critical Noise for Quasi-Consensus

As we know, the conventional consensus/synchronization concept signifies that the states of all agents are exactly the same, and this concept has been well studied in noise-free opinion dynamics and multi-agent systems. However, if considering a system affected by noise, the strict consensus/synchronization behavior may not be reached.

According to the HK model (3.1), once all the opinions locate within the confidence threshold, the agents will form a cluster and share the same average opinion in the next time. To deal with the noisy situation, we define the quasi-consensus of system (3.3) ~ (3.4). Denote

$$d_V(t) = \max_{i,j \in V} |x_i(t) - x_j(t)| \text{ and } d_V = \limsup_{t \to \infty} d_V(t)$$

Definition 3.1 (1) If $d_V \leqslant \varepsilon$, we say the system (3.3) ~ (3.4) will reach quasi-consensus.

(2) If $\mathbb{P}\{d_V \leqslant \varepsilon\} = 1$, we say almost surely (a. s.) the system (3.3) ~ (3.4) will V reach quasi-consensus.

(3) If $\mathbb{P}\{d_V \leqslant \varepsilon\} = 0$, we say a. s. the system (3.3) ~ (3.4) cannot reach quasi-consensus.

(4) Let $T = \min\{t : d_V(t') \leqslant \varepsilon \text{ for all } t' \geqslant t\}$. If $\mathbb{P}\{T < \infty\} = 1$, we say a. s. the system (3.3) ~ (3.4) reaches quasi-consensus in finite time.

From Definition 3.1, if the noisy HK dynamics (3.3) ~ (3.4) reaches quasi-consensus, all agents finally become neighbors of one another, and share a common average value of neighbors' opinions. Thus, from (3.4) the maximum difference between agents' opinions is not expected to exceed 2δ, where $\delta = \sup_{i \in v}^{t>0} |\xi_i(t)|$ can be viewed as noise strength (refer to Theorem 3.2). When δ is small, the opinions of the agents slightly differ and present a clear consensus picture. Small noise is more common in a large society, and quasi-consensus in this situation possesses increased practical significance because people rarely hold a completely identical opinion in a community even though they agree with one another in principle.

For simplicity, we first present the result of a quasi-consensus for independent and identically distributed (i. i. d.) noise. This noise condition can be relaxed in Sec. 3.3 and 3.4, where a sufficient condition and a necessary condition for quasi-consensus are provided.

Theorem 3.1 Suppose the noises $\{\xi_i(t)\}_{i \in v, t \geqslant 1}$ are zero-mean and non-degenerate random variables with independent and identical distribution, and $\mathbf{E}\xi_1^2(1) < \infty$. Let $x(0) \in [0,1]^n$ be arbitrarily given, then

(1) If $\mathbb{P}\left\{|\xi_1(1)| \leqslant \frac{\varepsilon}{2}\right\} = 1$, then a. s. the system (3.3) ~ (3.4) will reach quasi-consensus in finite time.

(2) Let the confidence threshold $\varepsilon \in (0, 1/3]$, if $\mathbb{P}\left\{\xi_1(1) > \frac{\varepsilon}{2}\right\} = 0$ and $\mathbb{P}\left\{\xi_1(1) < -\frac{\varepsilon}{2}\right\} > 0$, then a. s. the system (3.3) ~ (3.4) cannot reach quasi-consensus.

Proof Part (1) follows from the following Theorem 3.2, since for the zero-mean and non-degenerate i. i. d. random variables, there exist constants $a, p \in (0,1)$ such that $\mathbb{P}\{\xi_i(t) \geqslant a\} \geqslant p$ and $\mathbb{P}\{\xi_i(t) \leqslant -a\} \geqslant p$. Also, part (2) can be obtained directly from Theorem 3.3, where the necessary condition is obtained for the general independent noises. □

Remark 3.1 Theorem 3.1 states that $\varepsilon/2$ is the critical strength for random noise to induce the quasi-consensus of HK opinion dynamics: (1) when noise strength is not larger than $\varepsilon/2$, the noisy HK model (3.3) ~ (3.4) can a. s. achieve quasi-consensus in finite time; and (2) if noise strength has a positive probability of being larger than $\varepsilon/2$, the noisy HK model a. s. cannot attain quasi-consensus.

Remark 3.2 The condition $\varepsilon \in (0, 1/3]$ in Theorem 3.1 (2) is a conservative choice following Theorem 3.3 to facilitate the analysis when we only assume the noises to be independent. This condition $\varepsilon \in (0,1/3]$ can be extended to $\varepsilon \in (0,1)$. Meanwhile, i. i. d. noise can be relaxed to noise with a positive joint probability density.

Theorem 3.1 implies that under small noise, the opinions in a homogeneous group can reach an approximate consensus after a long time; conversely, under large noises, which represent an unstable social environment, infinite times must exist such that the opinions produce a large difference. These results fit social reality intuitively. Moreover, the results somehow imply that the free flow of information in social networks enhances the opinion agreement and provide insight for the design of opinion intervention strategy to reduce the disagreement by intentionally injecting specific noises into a divisive group.

3.3 Sufficient Condition for Quasi-Consensus

The following result provides a sufficient condition for the quasi-consensus with general independent noises.

Theorem 3.2 Suppose $\varepsilon \in (0,1]$, $\{\xi_i(t), i \in V, t \geqslant 1\}$ are independent and satisfy (1) $\mathbb{P}\{|\xi_i(t)| \leqslant \delta\} = 1$ with $0 < \delta \leqslant \varepsilon/2$; (2) there exist constants $a \in (0,\delta)$, $p \in (0,1)$ such that $\mathbb{P}\{\xi_i(t) \geqslant a\} \geqslant p$ and $\mathbb{P}\{\xi_i(t) \leqslant -a\} \geqslant p$. Then, for any initial state $x(0) \in [0,1]^n$, the system (3.3) ~ (3.4) will a. s. reach quasi-consensus in finite time and $d_V \leqslant 2\delta$ a. s..

To prove this result we need introduce some lemmas first. The following result is quite straightforward, which was used in[2].

Lemma 3.1 Suppose $\{z_i,\ i = 1, 2, \cdots\}$ is a nonnegative nondecreasing (nonincreasing) sequence. Then for any $s \geqslant 0$, the sequence, $\left\{g_s(k) = \frac{1}{k}\sum_{i=s+1}^{s+k} z_i,\ k \geqslant 1\right\}$ is monotonically nondecreasing (nonincreasing) for k.

In what follows, the ever appearing time symbols t (or T, etc.) all refer to the random variables $t(\omega)$ (or $T(\omega)$, etc.) on the probability space $(\Omega, \mathcal{F}, \mathbb{P})$, and will be still written as t (or T, etc.) for simplicity.

Lemma 3.2 For the system (3.3) ~ (3.4) with conditions of Theorem 3.2 (1), if a. s. there exists a finite time $0 \leqslant T < \infty$ such that $d_V(T) \leqslant \varepsilon$, then we have a. s. $d_V \leqslant 2\delta$.

Proof Denote $\tilde{x}_i(t) = \frac{1}{|N(i,x(t))|}\sum_{j\in N(i,x(t))} x_j(t),\ t\geqslant 0$, and this denotation remains valid for the rest of the context. If $d_V(T)\leqslant\varepsilon$, by (3.2) we have

$$\tilde{x}_i(T) = \frac{1}{n}\sum_{j=1}^{n} x_j(T),\ i\in V \tag{3.5}$$

Since $|\xi_i(t)|\leqslant\delta$ a. s., we obtain a. s.

$$\begin{aligned} d_V(T+1) &= \max_{1\leqslant i,j\leqslant n} |x_i(T+1)-x_j(T+1)| \\ &\leqslant \max_{1\leqslant i,j\leqslant n} (|\xi_i(T+1)|+|\xi_j(T+1)|)\leqslant 2\delta\leqslant\varepsilon \end{aligned} \tag{3.6}$$

Using (3.5) and (3.6) repeatedly, we get $d_V\leqslant 2\delta$ a. s. □

Lemma 3.2 indicates that, once all the opinions locate within the confidence threshold, the noises with strength no more than $\varepsilon/2$ cannot separate them any more and the system (3.3) ~ (3.4) reaches quasi-consensus.

Proof of Theorem 3.2 By Lemma 3.2. we only need to prove the system will a. s. reach quasi-consensus in finite time. Define for $t\geqslant 0$,

$$x_{\max}(t) = \max_{i\in V} x_i(t) \text{ and } x_{\min}(t) = \min_{i\in V} x_i(t)$$

Given any initial condition $x(0)$, if $d_V(0) = x_{\max}(0) - x_{\min}(0)\leqslant\varepsilon$, by Lemma 3.2, (3.3) ~ (3.4) is quasi-consensus. Otherwise, $d_V(0)>0$. Note that there exist constants $0<a<\varepsilon/2$ and $0<p<1$ such that, for $t\geqslant 1$, $k\in V$,

$$\mathbb{P}\{\xi_k(t)\geqslant a\}\geqslant p \text{ and } \mathbb{P}\{\xi_k(t)\leqslant -a\}\geqslant p \tag{3.7}$$

For all $t\geqslant 0$, let

$$\underline{V}(t) = \left\{ i\in V \,\middle|\, x_{\min}(t)\leqslant\tilde{x}_i(t)\leqslant x_{\min}(t)+\frac{d_V(t)}{2} \right\}$$

and

$$\overline{V}(t) = \left\{ j\in V \,\middle|\, x_{\min}(t)+\frac{d_V(t)}{2} < \tilde{x}_j(t)\leqslant x_{\max}(t) \right\}$$

Then for $i\in\underline{V}(0)$, $j\in\overline{V}(0)$, we have

$$\begin{aligned} &\mathbb{P}\{x_i(1)\geqslant\tilde{x}_i(0)+a\}\geqslant\mathbb{P}\{\xi_i(1)\geqslant a\}\geqslant p \\ &\mathbb{P}\{x_j(1)\leqslant\tilde{x}_j(0)-a\}\geqslant\mathbb{P}\{\xi_j(1)\leqslant -a\}\geqslant p \end{aligned}$$

Thus, by Lemma 3.1 and (3.7), we have

$$\mathbb{P}\{d_V(1)\leqslant d_V(0)-2a\}\geqslant\mathbb{P}\{\xi_i(1)\geqslant a,\ \xi_j(1)\leqslant -a,\ i\in \underline{V}(0),\ j\in \overline{V}(0)\}\geqslant p^n>0 \tag{3.8}$$

According to Lemma 3.2, (3.3) ~ (3.4) will reach quasi-consensus once $d_V(t)\leqslant\varepsilon$ at any time t. Let $L=\left\lceil\frac{1-\varepsilon}{2a}\right\rceil$, then following the similar argument of (3.8), we have

$$\begin{aligned}\mathbb{P}\{d_V(L+1)\leqslant\varepsilon\}&\geqslant\mathbb{P}\left\{\bigcap_{t=0}^{L}\{d_V(t+1)\leqslant d_V(t)-2a\}\right\}\\&\geqslant\mathbb{P}\left\{\bigcap_{t=0}^{L}\{\xi_i(t+1)\geqslant a,\ \xi_j(t+1)\leqslant -a,\ i\in\underline{V}(t),\ j\in\overline{V}(t)\}\right\}\end{aligned} \tag{3.9}$$

Denote events $E(0)=\Omega$, $E(t+1)=\{\omega:\xi_i(t+1)\geqslant a,\ \xi_j(t+1)\leqslant -a,\ i\in\underline{V}(t),\ j\in\overline{V}(t)\}$, $t\geqslant0$, and note that $\underline{V}(t)\cap\overline{V}(t)=\varnothing$, then by independence and following the discussion of (3.8) we have

$$\mathbb{P}\left\{E(t+1)\,\middle|\,\bigcap_{s=0}^{t}E(s)\right\}\geqslant p^n>0$$

hence by (3.9), we have

$$\begin{aligned}\mathbb{P}\{d_V(L+1)\leqslant\varepsilon\}&=\mathbb{P}\left\{\bigcap_{t=0}^{L}E(t+1)\right\}\\&=\prod_{0\leqslant t\leqslant L}\mathbb{P}\left\{E(t+1)\,\middle|\,\bigcap_{s=0}^{t}E(s)\right\}\cdot\mathbb{P}\{E(0)\}\\&\geqslant p^{n(L+1)}>0\end{aligned} \tag{3.10}$$

Define $U(L)=\{\omega:(3.3)\sim(3.4)$ does not reach quasi-consensus in period $L\}$ and $U=\{\omega:(3.3)\sim(3.4)$ does not reach quasi-consensus in finite time$\}$. By (3.10), $\mathbb{P}\{U(L)\}\leqslant1-p^{n(L+1)}<1$. Since $x(0)$ is arbitrarily given in $[0,1]^n$, following the procedure of (3.10), we can obtain that

$$\mathbb{P}\{U(mL)\mid U((m-1)L)\}\leqslant1-p^{n(L+1)},\ m>1$$

Then

$$\begin{aligned}\mathbb{P}\{U\}&=\mathbb{P}\left\{\bigcap_{m=1}^{\infty}U(mL)\right\}=\lim_{m\to\infty}\mathbb{P}\{U(mL)\}\\&=\lim_{m\to\infty}\prod_{k=1}^{m-1}\mathbb{P}\{U((k+1)L)\mid U(kL)\}\cdot\mathbb{P}\{U(L)\}\end{aligned}$$

$$\leqslant \lim_{m\to\infty}(1-p^{n(L+1)})^m = 0$$

and hence,

$$\mathbb{P}\{(3.3)\sim(3.4) \text{ can reach quasi-consensus in finite time}\} = 1-\mathbb{P}\{U\} = 1 \quad \square$$

Fig. 3.1 shows that noise-free HK dynamics cannot reach consensus for some initial opinions and confidence thresholds. Theorem 3.2 shows that even tiny noises can eventually cause collective opinions to merge. The consensus behavior of multi-agent systems with or without noise can be ensured by graph connectivity in some sense during the evolution[9, 13, 14]. Theorem 3.2 implies that persistent noises can produce the connectedness of HK opinion dynamics even when noise is rather weak.

3.4 Necessary Condition for Quasi-Consensus

Here we show that when the noise strength is larger than the critical half confidence threshold, the opinions almost surely cannot reach quasi-consensus.

Theorem 3.3 Let $x(0)\in[0,1]^n$, $\varepsilon\in\left(0,\frac{1}{3}\right]$. Assume the zero-mean noises $\{\xi_i(t), i\in V, t\geqslant 1\}$ are i. i. d. with $\mathbf{E}\xi_1^2(1)<\infty$ or independent with $\sup_{i,t}|\xi_i(t)|<\infty$, a. s. If there exists a lower bound $q>0$ such that $\mathbb{P}\left\{\xi_i(t)>\frac{\varepsilon}{2}\right\}\geqslant q$ and $\mathbb{P}\left\{\xi_i(t)<-\frac{\varepsilon}{2}\right\}\geqslant q$, then a. s. the system (3.3) ~ (3.4) cannot reach quasi-consensus.

Prior to the proof of Theorem 3.3, we first check how the noisy opinions behave after they reach quasi-consensus. Intuitively, the clustered opinions fluctuate slightly when the noise is weak, but the subsequent lemma shows this seemingly slight fluctuation could only maintain in a relatively short time; as time goes long enough, the synchronized noisy opinions can cross the entire interval $[0,1]$ with infinite times.

For an event sequence $\{B_m, m\geqslant 1\}$, let $\{B_m, \text{ i. o.}\}$ be the set of $\bigcap_{m=1}^{\infty}\bigcup_{s=m}^{\infty}B_s$, where i. o. is the abbreviation of "infinitely often". Then we have the following lemma.

Lemma 3.3 Suppose the non-degenerate zero-mean random noises $\{\xi_i(t), i\in V, t\geqslant 1\}$ are i. i. d. with $\mathbf{E}\xi_1^2(1)<\infty$ or independent with $\inf_{i,t}\mathbf{E}\xi_i^2(t)>0$ and $\sup_{i,t}|\xi_i(t)|<\infty$, a. s. $\}$. If a. s. there exists a finite time $0\leqslant T<\infty$ such that $d_V(t)\leqslant\varepsilon$ for $t\geqslant T$, then it a. s. occurs $x_i(t)=0$ i. o. and $x_i(t)=1$ i. o. for $i\in V$.

Proof We only need to prove the case when the noise is independent, and the

i. i. d. case can be obtained similarly. Without loss of generality, we assume $T=0$ a. s. and only need to prove $\mathbb{P}\{x_i(t)=1, \text{ i. o.}\}=1$. Consider another modified noisy HK model as follows

$$y_i(t+1) = \frac{1}{|N(i, y(t))|}\sum_{j\in N(i,y(t))} y_j(t) + \xi_i(t+1) \tag{3.11}$$

for $i\in V$, $t\geqslant 0$. According to the proof of Lemma 3.2, we have for any $i\in V$, $t>0$,

$$\begin{aligned} y_i(t+1) &= \frac{1}{n}\sum_{j=1}^{n} y_j(t) + \xi_i(t+1) \\ &= \frac{1}{n}\sum_{j=1}^{n}\left(\frac{1}{n}\sum_{k=1}^{n} y_k(t-1) + \xi_j(t)\right) + \xi_i(t+1) \\ &= \frac{1}{n}\sum_{j=1}^{n} y_j(t-1) + \frac{1}{n}\sum_{j=1}^{n}\xi_j(t) + \xi_i(t+1) \\ &= \frac{1}{n}\sum_{j=1}^{n} y_j(0) + \sum_{k=1}^{t}\frac{1}{n}\sum_{j=1}^{n}\xi_j(k) + \xi_i(t+1) \end{aligned} \tag{3.12}$$

Denote $\eta_k = \frac{1}{n}\sum_{j=1}^{n}\xi_j(k)$, $k\geqslant 1$, and suppose $\sup_{i,t}|\xi_i(t)|\leqslant d<\infty$, a. s. for some $d>0$, and $\inf_{i,t}\mathbf{E}\xi_i^2(t)\geqslant c>0$. Then $\{\eta_k, k\geqslant 1\}$ are mutually independent with $|\eta_k|\leqslant d$ a. s., $\mathbf{E}\eta_k=0$, and $\mathbf{E}\eta_k^2 = \frac{1}{n^2}\sum_{j=1}^{n}\mathbf{E}\xi_j^2(k)\geqslant\frac{1}{n}c$, which implies $\mathbf{E}\eta_k^2\in\left[\frac{c}{n}, \frac{d^2}{n}\right]$. Let $S_t=\sum_{k=1}^{t}\eta_k$, then $s_t^2=\mathbf{E}S_t^2=\sum_{k=1}^{t}\mathbf{E}\eta_k^2\in\left[t\frac{c}{n}, t\frac{d^2}{n}\right]$. Hence, $\lim_{t\to\infty}s_t^2=+\infty$ and

$$\frac{d}{(s_t/\sqrt{\log\log s_t})}\leqslant\frac{d\sqrt{\log\log(\sqrt{td^2/n})}}{\sqrt{tc/n}}\to 0$$

As $t\to\infty$. According to Lemma 2.3

$$\mathbb{P}\left\{\overline{\lim_{t\to\infty}}S_t=+\infty\right\}=\mathbb{P}\left\{\underline{\lim_{t\to\infty}}S_t=-\infty\right\}=1 \tag{3.13}$$

Again with (12), we have for any $y_i(0)\in[0,1]$

$$\mathbb{P}\left\{\overline{\lim_{t\to\infty}}y_i(t)=+\infty\right\}=\mathbb{P}\left\{\underline{\lim_{t\to\infty}}y_i(t)=-\infty\right\}=1 \tag{3.14}$$

As a result

$$\begin{aligned}\mathbb{P}\{x_i(t)=1,\ i.\ o.\} &= \mathbb{P}\left\{\bigcap_{m=1}^{\infty}\bigcup_{s=m}^{\infty}\{x_i(s)=1\}\right\}\\ &=1-\mathbb{P}\left\{\lim_{m\to\infty}\bigcap_{s=m}^{\infty}\{x_i(s)<1\}\right\}\\ &=1-\lim_{m\to\infty}\mathbb{P}\left\{\bigcap_{s=m}^{\infty}\{x_i(s)<1\}\right\}\end{aligned} \tag{3.15}$$

where the last equality holds since $\left\{\bigcap_{s=m}^{\infty}\{x_i(s)<1\},\ m\geqslant 1\right\}$ is increasing. For each $m\geqslant 1$, if $y_i(m)=x_i(m)$, $i\in V$, Lemmas 3.1 and 3.2 along with (3.3) and (3.12) imply that $y_i(t)(\omega)\leqslant x_i(t)(\omega)$, $t\geqslant m$ for all $\omega\in\bigcap_{s=m}^{\infty}\{x_i(s)<1\}$ and hence, $\bigcap_{s=m}^{\infty}\{x_i(s)<1\}\subset\bigcap_{s=m}^{\infty}\{y_i(s)<1\}$. By (3.15)

$$\begin{aligned}\mathbb{P}\{x_i(t)=1,\ \text{i. o.}\} &=1-\lim_{m\to\infty}\mathbb{P}\left\{\bigcap_{s=m}^{\infty}\{x_i(s)<1\}\right\}\\ &\geqslant 1-\lim_{m\to\infty}\mathbb{P}\left\{\bigcap_{s=m}^{\infty}\{y_i(s)<1\},\ y_i(m)=x_i(m)\right\}\end{aligned} \tag{3.16}$$

From (3.14), we have that, for $m\geqslant 1$

$$\mathbb{P}\left\{\bigcap_{s=m}^{\infty}\{y_i(s)<1\},\ y_i(m)=x_i(m)\right\}=0 \tag{3.17}$$

Thus, $\mathbb{P}\{x_i(t)=1,\ \text{i. o.}\}=1$ by (3.16) and (3.17). □

Proof of Theorem 3.3 We only need to prove the independent case, while the i. i. d. case can be obtained similarly. For the independent case, we only need to prove that, for any $T\geqslant 0$, there exists $t\geqslant T$ a. s. such that $d_V(t)>\varepsilon$ when $\varepsilon\leqslant\frac{1}{3}$, that is,

$$\mathbb{P}\left\{\bigcup_{T=0}^{\infty}\{d_V(t)\leqslant\varepsilon,\ t\geqslant T\}\right\}=0$$

Given arbitrary $T<\infty$, it is not hard to see that, on $\{d_V(t)\leqslant\varepsilon,\ t\geqslant T\}$, we have

$$\begin{aligned}&\mathbb{P}\left\{d_V(t)\leqslant\varepsilon,\ \bigcap_{k=1}^{\infty}\bigcup_{i\in V}\{\xi_i(t_{m_k})>\varepsilon\}\right\}\\ &\leqslant\mathbb{P}\left\{d_V(t)\leqslant\varepsilon,\ \bigcap_{k=1}^{\infty}\bigcup_{i\in V}\bigcap_{j\in V}\{\xi_i(t_{m_k})>\varepsilon,\ \xi_j(t_{m_k})\geqslant 0\}\right\}\\ &\leqslant\mathbb{P}\left\{\bigcap_{k=1}^{\infty}\bigcap_{j\in V}|\{\xi_j(t_{m_k})\geqslant 0\}\right\}\\ &\leqslant\prod_{k=1}^{\infty}\prod_{j\in V}\left(1-\mathbb{P}\left\{\xi_j(t_{m_k})<-\frac{\varepsilon}{2}\right\}\right)\leqslant\lim_{k\to\infty}(1-q)^{nk}=0\end{aligned} \tag{3.18}$$

Similarly

$$\mathbb{P}\left\{d_V(t)\leqslant\varepsilon,\ \bigcap_{k=1}^{\infty}\bigcup_{i\in V}\{\xi_i(t_{m_k})<-\varepsilon\}\right\}=0 \tag{3.19}$$

(3.18) and (3.19) imply that a. s. there are at most only finite times that the noise strength exceeds ε when the system reaches quasi-consensus. Since $\varepsilon\leqslant\frac{1}{3}$, by Lemma 3.3, there a. s exists a finite stopping time sequence $T\leqslant T_1\leqslant T_2<\cdots$ such that $x_i(T_j)\in[\varepsilon,1-\varepsilon]$, $i\in V$, $j\geqslant 1$.

Denote $M(t)=\{u\in V\mid x_u(t)\leqslant x_v(t),\ v\in V\}$, $t\geqslant 0$ and note that

$$\mathbb{P}\left\{\xi_i(t)>\frac{\varepsilon}{2}\right\}\geqslant q \text{ and } \mathbb{P}\left\{\xi_i(t)<-\frac{\varepsilon}{2}\right\}\geqslant q \tag{3.20}$$

for $i\in V$, $t>0$. Then, it follows from the independence of $\{\xi_i(t),\ i\in V,\ t>0\}$ that, for $\alpha\in M(t)$, $\beta\in V-M(t)$, $j\geqslant 1$

$$\begin{aligned}\mathbb{P}\{d_V(T_j+1)>\varepsilon\}&\geqslant\mathbb{P}\{\xi_\alpha(T_j+1)<-\varepsilon/2,\xi_\beta(T_j+1)>\varepsilon/2\}\\&=\sum_{s=0}^{\infty}\mathbb{P}\{\xi_\alpha(s+1)<-\varepsilon/2,\ \xi_\beta(s+1)>\varepsilon/2\mid T_j=s\}\cdot\mathbb{P}\{T_j=s\}\\&=\sum_{s=0}^{\infty}\mathbb{P}\{\xi_\alpha(s+1)<-\varepsilon/2,\ \xi_\beta(s+1)>\varepsilon/2\}\cdot\mathbb{P}\{T_j=s\}\\&\geqslant q^n>0\end{aligned} \tag{3.21}$$

Denote $E_0=\Omega$ and $E_j=\{\omega:d_V(t)\leqslant\varepsilon,\ t\in[T_j+1,T_{j+1}]\}$ for $j\geqslant 1$. Since $E_j\in\sigma(T_j+1,\cdots)$ is independent with $\sigma(T_j)$, then

$$\mathbb{P}\left\{E_j\,\Big|\,\bigcap_{l<j}E_l\right\}\leqslant\mathbb{P}\{d_V(T_j+1)\leqslant\varepsilon\}\leqslant 1-q^n<1$$

It follows that

$$\begin{aligned}\mathbb{P}\{d_V(t)\leqslant\varepsilon,\ t\geqslant T\}&\leqslant\mathbb{P}\left\{\bigcap_{j\geqslant 1}E_j\right\}\\&=\lim_{m\to\infty}\prod_{j=1}^{m}\mathbb{P}\left\{E_j\,\Big|\,\bigcap_{l<j}E_l\right\}\cdot\mathbb{P}\{E_0\}\\&\leqslant\lim_{m\to\infty}(1-q^n)^m=0\end{aligned}$$

This completes the proof. □

3.5 Simulations

In this section, we provide simulations for the system (3.3) ~ (3.4) to verify the theoretical results.

Notably, when noise strength is below the critical half confidence threshold but relatively strong to the opinions, the noisy opinions can reach quasi-consensus but fluctuate sharply, drive by noises. Therefore, for a clear illustration, we simply focus on the case when noise strength is weak, which shows more clearly how consensus can be achieved. Given $n = 20$, $\varepsilon = 0.2$, and the initial opinion values uniformly distributed on $[0,1]$, Fig. 3.1 shows the evolution of noise-free opinions with 3 clusters formed. Let $\xi_i(t)$, $i \in V$, $t \geqslant 1$ be the noises uniformly distributed on $[-\delta, \delta]$. Consider Theorem 3.1 (1) and let $\delta = 0.1\varepsilon$, then all opinions merge into 1 group as shown in Fig. 3.2.

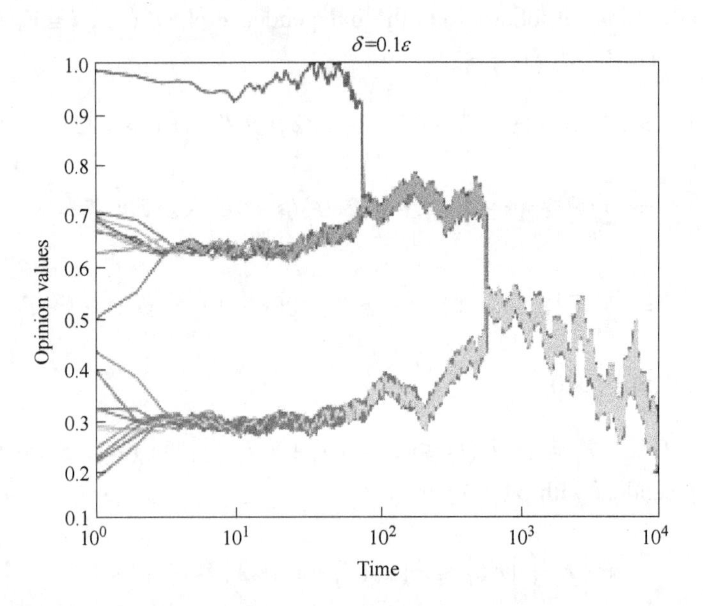

Fig. 3.2　The opinion evolution of (3.3) ~ (3.4) when all agents are noise infected with $\delta = 0.1\varepsilon$

(The initial states and confidence threshold are the same as those in Fig. 3.1. It can be seen that the opinions get merged and (3.3) ~ (3.4) reaches quasi-consensus)

We then show the divergence of opinions when noise strength exceeds the critical value. For simplicity, we set the initial opinions to be identical, and show that the noise with a strength larger than 0.5ε can separate the opinions at some moment. Also, when the confidence threshold is high, the noise strength becomes too strong that the noisy opinions fluctuate sharply, though they may be separated. Therefore, we use a low confidence threshold for a clear illustration. Let $n = 10$, $\varepsilon = 0.01$, all initial opinion values be 0.5, and $\delta = 0.6\varepsilon > 0.5\varepsilon$ by Theorem 3.1 (2); then the gathered opinions divide at some moment in Fig. 3.3. Some similar simulation results for synchronized and divergent phenomena of noisy opinions can be also found in [3, 4, 7].

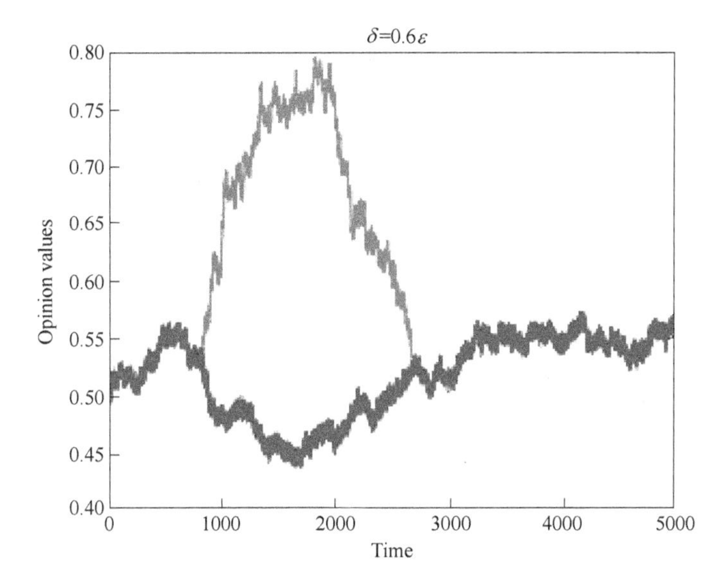

Fig. 3.3 The opinion evolution of (3.3) ~ (3.4) when all agents are noiseinfected with $\delta = 0.6\varepsilon$ (It illustrates that some opinions get separated from the others at some moment though they are gathered at the beginning)

3.6 Notes

The agreement/disagreement analysis of opinion dynamics has become increasingly important in recent years. In this chapter, it is provided a strict analysis for the proposed noisy HK opinion model and proved that the random noise could almost surely make the HK dynamics achieve "consensus" (quasi-consensus), while the critical noise strength was found. Although the strict mathematical investigation of the noisy confidence-based opinion models is very hard and rare, we have developed a theoretical method for the noisy HK model, which may be later extended to other confidence-based opinion dynamics. In addition, these results reveal a richer mechanism of opinion evolution, and also provide further insights into designing noise intervention strategy to eliminate social disagreement.

References

[1] Hegselmann R, Krause U. Opinion dynamics and bounded confidence models, analysis, and simulation[J]. Journal of Artificial Societies and Social Simulation, 2002, 5 (3): 1 ~33.

[2] Krause U. A discrete nonlinear and non-automonous model of consensus formation. Communications in Difference Equations, Amsterdam: Gordon and Breach Publisher, 2000, 227 ~238.

[3] Mäs M, Flache A D. Helbing, Individualization as driving force of clustering phenomena in humans[J]. PLoS Computational Biology, 2010, 6 (10), e1000959.

[4] Pineda M, Toral R, Hernndez-Garca E. Diffusing opinions in bounded confidence processes[J]. The European Physical Journal D, 2011, 62 (1): 109 ~117.

[5] Grauwin S, Jensen P. Opinion group formation and dynamics: Structures that last from nonlasting entities[J]. Physical Review E, 2012, 85 (6), 066113.

[6] Carro A, Toral R, San Miguel M. The role of noise and initial conditions in the asymptotic solution of a bounded confidence, continuous-opinion model[J]. Journal of Statistical Physics, 2013, 151 (1 ~2): 131 ~149.

[7] Pineda M, Toral R, Hernndez-Garca E. The noisy Hegselmann-Krause model for opinion dynamics[J]. The European Physical Journal B, 2013, 86 (12): 1 ~10.

[8] Huang M, Manton J. Opinion dynamics with noisy information[C]. Proceedings of the IEEE Conference on Decision and Control, 2013, 3445 ~3450.

[9] Wang L, Guo L. Robust consensus and soft control of multi-agent systems with noises[J]. Journal of Systems Science and Complexity, 2008, 21 (3): 406 ~415.

[10] Kar S, Moura J F. Distributed consensus algorithms in sensor networks with imperfect communications: Link failures and channel noise[J]. IEEE Transactions on Signal Processing, 2009, 57 (1): 355 ~369.

[11] Amelina A, Fradkov A, Jiang Y, Vergados D. Approximate consensus in stochastic networks with application to load balancing[J]. IEEE Transactions on Information Theory, 2015, 61 (4): 1739 ~1752.

[12] Shang Y. L1 group consensus of multi-agent systems with switching topologies and stochastic inputs[J]. Physics Letters A, 2013, 377: 1582 ~1586.

[13] Touri B, Langbort C. On endogenous random consensus and averaging dynamics[J]. IEEE Transactions on Control of Network Systems, 2015, 1 (3): 241 ~248.

[14] Jadbabaie A, Lin J, Morse A S. Coordination of groups of mobile autonomous agents using nearest neighbor rules[J]. IEEE Transactions on Automatic Control, 2003, 48 (6): 988 ~1001.

4 Homogeneous HK Model in Unbounded Space

In Chapter 3, we established a theoretical analysis of noise-induced synchronization based on HK model of opinion dynamics. In the HK model, each agent possesses a bounded confidence and updates its opinion value by averaging the opinions of its neighbors who are located within its confidence region. In spite of its seeming simplicity, the HK model captures a quite fundamental local rule of evolution which is embodied ubiquitously in self-organizing systems, such as the Boid and Vicsek models, and has been largely explored in its deterministic version[1~4]. The analysis of noisy HK models in previous studies was subject to an assumption that all agents' opinions were limited to a bounded interval[5~9]. In particular, the boundedness assumption was crucial to the proof of noise-induced synchronization of HK dynamics in Chapter 3. When the state-space is bounded, the system has a uniform positive probability to reach quasi-synchronization in a finite period from any initial state. However, when the state of a noisy HK model is allowed to exist in the full space, the system has no uniform positive probability to attain quasi-synchronization in a finite period from any initial state, leading thus to invalidation of the existing methods.

In this chapter, via exploring the topological property of a noisy HK system in full state-space, and using the theory of independent stopping time, we show that for any initial state, the system will reach a state whose neighboring graph consists of all complete subgraphs, with a uniform positive probability in a finite period. Additionally, given any initial state whose neighboring graph consists of all complete subgraphs, we prove that the system will achieve quasi-synchronization with a uniform positive probability in an almost surely (a. s.) finite stopping time. Combining the two conclusions leads us to the final answer. Importantly, we wish to stress here that finding the uniform positive probability in a finite stopping time is essentially a new skill which may extend the idea of "joint connectivity in a finite period" to the "joint connectivity in a finite stopping time" in the consensus of multi-agent systems.

Besides this novel mathematical achievement, another highlighting contribution of this chapter is its physical significance in providing a theoretical interpretation of the noise-induced synchronization of self-organizing systems. Though the HK model with bounded state-space performs generally well in mimicking opinion behavior, the imposed assumption of boundedness of the state-space is undesirable in physical systems and largely limits its potential application in representing an elementary self-organizing system.

4.1 Model and Definitions

Denot $V=\{1,2,\cdots,n\}$ as the set of n agents, $x_i(t)\in(-\infty,\infty)$, $i\in V$, $t\geqslant 0$ as the state of agent i at time t. The update rule of HK dynamics then takes:

$$x_i(t+1)=\frac{1}{|N_i(x(t))|}\sum_{j\in N_i(x(t))}x_j(t)+\xi_i(t+1),\ i\in V \tag{4.1}$$

Where

$$N_i(x(t))=\{j\in V\mid |x_j(t)-x_i(t)|\leqslant\varepsilon\} \tag{4.2}$$

is the neighbor set of i at t, $\varepsilon>0$ represents the confidence threshold of the agents and $\xi_i(t)$, $i\in V$, $t\geqslant 1$ is noise. Here, $|\cdot|$ can be the cardinal number of a set or the absolute value of a real number.

In Chapter 3, the state-space is assumed to be bounded, i. e. $x_i(t)\in[0,1]$, $i\in V$, $t\geqslant 0$. If there is no noise, it is proved that for any given initial opinion value $x(0)\in[0,1]^n$, the evolutionary opinion values $x(t)$, $t\geqslant 0$ of the noise-free HK model cannot exceed the initial boundary opinions. However, in the presence of noise, mathematically, the evolutionary opinion values can be driven to run outside the initial boundary opinions, and even outside the opinion space $[0,1]$. In [5], to limit the noisy opinion values in $[0,1]$, it forcibly assumes that $x_i(t+1)=0$ or 1 when $\frac{1}{|N_i(x(t))|}\sum_{j\in N_i(x(t))}x_j(t)+\xi_i(t+1)$ is less than 0 or larger than 1. In model (4.1), this assumption is cancelled and the state-space is allowed to be unbounded.

To proceed, some preliminary definitions are first needed.

Definition 4.1 Let $\mathcal{G}_V(t)=\{V,\varepsilon(t)\}$ be the graph of V at time t, and $(i,j)\in\varepsilon(t)$ if and only if $|x_i(t)-x_j(t)|\leqslant\varepsilon$. A graph $\mathcal{G}_V(t)$ is called a complete graph if and only if $(i,j)\in\varepsilon(t)$ for any $i,j\in V$; and $\mathcal{G}_V(t)$ is called a connected graph if and only if for any $i\neq j$, there are edges (i,i_1), (i_1,i_2), $\cdots$, (i_k,j) in $\varepsilon(t)$.

The definition of quasi-synchronization of the noisy model (4.1) ~ (4.2) is given similarly as quasi-consensus in Chapter 3.

Definition 4.2 Denote

$$d_V(t) = \max_{i,j \in V} |x_i(t) - x_j(t)| \text{ and } d_V = \limsup_{t \to \infty} d_V(t)$$

(1) If $d_V \leqslant \varepsilon$, we say the system (1) ~ (2) will reach quasi-synchronization.

(2) If $\mathbb{P}\{d_V \leqslant \varepsilon\} = 1$, we say almost surely (a. s.) the system (1) ~ (2) will reach quasi-synchronization.

(3) If $\mathbb{P}\{d_V \leqslant \varepsilon\} = 0$, we say a. s. the system (1) ~ (2) cannot reach quasi-synchronization.

(4) Let $T = \min\{t : d_V(t') \leqslant \varepsilon \text{ for all } t' \geqslant t\}$. If $\mathbb{P}\{T < \infty\} = 1$, we say a. s. the system (1) ~ (2) reaches quasi-synchronization in finite time.

4.2 Main Results

For simplicity, we first present the result for quasi-synchronization with independent and identically distributed (i. i. d.) noises, which can be directly derived from the two subsequent general results with independent noises.

Theorem 4.1 (Critical noise amplitude for quasi-synchronization of HK model with i. i. d. noise) Let $\{\xi_i(t)\}_{i \in V, t \geqslant 1}$ be non-degenerate random variables with independent and identical distribution, then for any $x(0) \in (-\infty, \infty)^n$ and $\varepsilon > 0$.

(1) If there exist constants $\delta_1, \delta_2 \in (-\infty, \infty)$ with $\delta_2 - \delta_1 = \varepsilon$ and $\mathbf{E}\xi_1(1) = \frac{\delta_2 + \delta_1}{2}$, $\mathbb{P}\{\delta_1 \leqslant \xi_1(1) \leqslant \delta_2\} = 1$, then a. s. the system (4.1) ~ (4.2) will reach quasi-synchronization in finite time.

(2) If $\mathbb{P}\{\delta_1 \leqslant \xi_1(1) \leqslant \delta_2\} < 1$ for any $\delta_2 - \delta_1 = \varepsilon$, then a. s. the system (4.1) ~ (4.2) cannot reach quasi-synchronization.

Conclusion (1) shows that if noise amplitude is no more than ε, the system will a. s. achieve quasi-synchronization in finite time; Conclusion (2) states that when noise amplitude has a positive probability to exceed ε, the system will not reach quasi-synchronization. This implies ε is the critical noise amplitude to induce a quasi-synchronization. Conclusions (1) and (2) can be directly derived from the following Theorems 4.2 and 4.3, which present sufficient and necessary conditions, respectively, for independent noises:

Theorem 4.2 (Sufficient condition for quasi-synchronization of HK model with independent noise) Let $\{\xi_i(t), i \in V, t \geqslant 1\}$ be independent random variables with $\mathbf{E}\xi_i(t) = C \in (-\infty, \infty)$ and satisfy: (1) $\mathbb{P}\{\delta_1 \leqslant \xi_i(t) \leqslant \delta_2\} = 1$ with $0 < \delta_2 - \delta_1 \leqslant \varepsilon$; (2) there exist constants $a \in \left(0, \frac{\delta_2 - \delta_1}{2}\right)$, $p \in (0,1)$ such that $\mathbb{P}\left\{\xi_i(t) \geqslant \frac{\delta_2 + \delta_1}{2} + a\right\} \geqslant p$ and

$\mathbb{P}\left\{\xi_i(t) \leqslant \frac{\delta_2+\delta_1}{2} - a\right\} \geqslant p$. Then, for any initial state $x(0) \in (-\infty, \infty)^n$ and $\varepsilon > 0$, the system (4.1) ~ (4.2) will a. s. reach quasi-synchronization in finite time and $d_V \leqslant \delta_2 - \delta_1$ a. s.

Proof of Theorem 4.1(1) Noting that for i. i. d. random variables $\{\xi_i(t), i \in V, t \geqslant 1\}$ with $\mathbf{E}\xi_1(1) = \frac{\delta_2+\delta_1}{2}$, $\mathbf{E}\xi_1^2(1) > 0$, there exist constants $a > 0$ and $0 < p \leqslant 1$, such that

$$\mathbb{P}\left\{x_i(t) - \frac{\delta_2+\delta_1}{2} > a\right\} \geqslant p,\ \mathbb{P}\left\{x_i(t) - \frac{\delta_2+\delta_1}{2} < -a\right\} \geqslant p$$

the conditions in Theorem 4.2 can be satisfied. □

To prove Theorem 4.2, some lemmas are need. In what follows, let $\xi(t) = \{\xi_i(t), i \in V\}$, and the ever appearing time symbols t(or T, etc.) all refer to the random variables $t(\omega)$(or $T(\omega)$, etc.) on the probability space $(\Omega, \mathcal{F}, \mathbb{P})$, and will be still written as t(or T, etc.) for simplicity.

Lemma 4.1 For the system (4.1) ~ (4.2) with conditions of Theorem 4.2 (1), if there exists a finite time $0 \leqslant T < \infty$ such that $d_V(T) \leqslant \varepsilon$, then we have $d_V(t) \leqslant \delta_2 - \delta_1$ for $t > T$.

Proof The proof is quite similar to that of Lemma 3.2.

The following lemma is key to obtain the uniformly positive probability of reaching quasi-synchronization from any initial states:

Lemma 4.2 For system (4.1) ~ (4.2) with conditions of Theorem 4.2 (1), if there exists a finite time T and disjoint subsets $V_k \subset V$, $k = 1, \cdots, m (1 \leqslant m \leqslant n)$ such that $d_{V_k}(T) \leqslant \varepsilon$, $1 \leqslant k \leqslant m$, and for $k_1 \neq k_2$, $V_{k_1} \cap V_{k_2} = \varnothing$, $|x_i(T) - x_j(T)| > \varepsilon$, $i \in V_{k_1}$, $j \in V_{k_2}$, then there exist constants $0 < p_0 \leqslant 1$, $L_0 > 0$ and a finite stopping time series T_i which is $\sigma\left(\xi\left((i-1)L_0 + T + \sum_{j=1}^{i-1} T_j + 1, \cdots\right)\right)$ - measurable, $i = 1, \cdots, m-1$ such that $\mathbb{P}\{d_V(T + (m-1)L_0 + T_1 + \cdots + T_{m-1}) \leqslant \delta_2 - \delta_1\} \geqslant p_0$.

Proof Without loss of generality, suppose $T = 0$ a. s. Then at the initial moment, the system forms m subgroups with complete graphs, and by (4.1), $d_{V_k}(1) \leqslant \delta_2 - \delta_1 \leqslant \varepsilon$, $k = 1, \cdots, m-1$. Before one subgroup enters the neighbor region of another, for each V_k, $1 \leqslant k \leqslant m$, we have

$$x_i(t+1) = \frac{1}{|V_k|}\sum_{j \in V_k} x_j(t) + \xi_i(t+1) = \frac{1}{|V_k|}\sum_{j \in V_k} x_j(0) + \sum_{l=1}^{t} \frac{\sum_{j \in V_k} \xi_j(l)}{|V_k|} + \xi_i(t+1) \tag{4.3}$$

Order the subgroups at any moment $t \geqslant 1$ by the state values as $1,2,\cdots,m$, and consider the subgroups $V_1(1)$ with smallest state values and $V_m(1)$ with the largest state values. For $t \geqslant 0, k = 1,\cdots,m$, let

$$y_k(t+1) = \frac{1}{|V_k|}\sum_{j\in V_k} x_j(0) + \sum_{l=1}^{t} \frac{\sum_{j\in V_k}\xi_j(l)}{|V_k|}$$

Then for $t \geqslant 1$

$$y_m(t) - y_1(t) = \frac{\sum_{j\in V_m} x_j(0)}{|V_m|} - \frac{\sum_{j\in V_1} x_j(0)}{|V_1|} + \sum_{l=1}^{t-1}\left(\frac{\sum_{j\in V_m}\xi_j(l)}{|V_m|} - \frac{\sum_{j\in V_1}\xi_j(l)}{|V_1|}\right)$$

Since $\xi(t) = \{\xi_i(t),\ i \in V\}$, $t \geqslant 1$ are independent, the σ - algebra $\sigma(\xi(t))$, $t \geqslant 1$ are independent. By Law of the Iterated Logarithm (Lemma 2.3), we have that

$$\begin{aligned} &\limsup_{t\to\infty}(y_m(t) - y_1(t)) = \infty, \text{ a. s.} \\ &\liminf_{t\to\infty}(y_m(t) - y_1(t)) = -\infty, \text{ a. s.} \end{aligned} \tag{4.4}$$

Notice that $\left|\frac{\sum_{j\in V_m}\xi_j(l)}{|V_m|} - \frac{\sum_{j\in V_1}\xi_j(l)}{|V_1|}\right| \leqslant \delta_2 - \delta_1 \leqslant \varepsilon$ a. s. , by (4.4), there exists a σ_t-time $0 \leqslant T_0 < \infty$ where $\sigma_t = \sigma(\xi(1),\cdots,\xi(t))$ that

$$0 < y_m(T_0) - y_1(T_0) \leqslant \varepsilon, \text{ a. s} \tag{4.5}$$

Combining (4.3) and (4.5), we obtain that there a. s. exists a σ_t-time $T_1 \leqslant T_0$ such that at T_1, at least two subgroups with complete graphs will for the first time enter the neighbor region of one another and become a new complete or connected graph. Denote $\tilde{x}_i(t) = |N(i,x(t))|^{-1}\sum_{j\in N(i,x(t))} x_j(t)$, $i \in V$, $t \geqslant 0$ and let $V_c(t)$ be the new emerging subgroups with connected but not complete graphs at t, then for $i \in V_c(t)$ design the following protocol:

$$\begin{cases} \xi_i(t+1) \in \left[\dfrac{\delta_2+\delta_1}{2} + a, \delta_2\right], \text{ if } \min\limits_{j\in V_c(t)} x_j(t) \leqslant x_i(t) \leqslant \min\limits_{j\in V_c(t)} x_j(t) + \dfrac{d_{V_c(t)}(t)}{2} \\ \xi_i(t+1) \in \left[\delta_1, \dfrac{\delta_2+\delta_1}{2} - a\right], \text{ if } \min\limits_{j\in V_c(t)} x_j(t) + \dfrac{d_{V_c(t)}(t)}{2} < x_i(t) \leqslant \max\limits_{i\in V_c(t)} x_j(t) \end{cases} \tag{4.6}$$

For all $V_c(t)$, by (4.1) and Lemma 3.1, we know that under protocol (4.6) the minimum state value of $V_c(t)$ increases by at least a, the maximum state value of $V_c(t)$ decreases by at least a, and $d_{V_c(t)}(t)$ reduces by at least $2a$ after each step. Moreover, we know that $d_{V_c(t)}(t) \leqslant n\varepsilon$, then under the protocol (4.6), there must exist a constant $L_0 \leqslant \left\lceil \frac{(n-1)\varepsilon}{2a} \right\rceil$ such that $d_{V_c}(T_1+L_0)(T_1+L_0) \leqslant \varepsilon$ a. s. (This also means protocol (4.6) occurs L_0 times). Since there exist $m \leqslant n$ subgroups with complete graphs at the initial moment, by following the above procedure, we obtain that under the protocol (4.6), the whole group V will a. s. form a complete graph in a finite time $\overline{T} \leqslant \sum_{1}^{m-1} T_j + (m-1)L_0$ where T_j is $\sigma\left(\xi\left((j-1)L_0 + \sum_{1}^{i-1} T_j + 1\right), \cdots\right)$-measurable, and during this process, protocol (4.6) occurs no more than $(m-1)L_0$ times. By independence of $\xi_i(t)$, $i \in V$, $t \geqslant 1$, we know that

$$\mathbb{P}\{\text{protocol } (4.6) \text{ occurs } (m-1)L_0 \text{ times}\} \geqslant p^{n(m-1)L_0} > 0$$

Let $p_0 = p^{n(n-1)L_0}$ and consider Lemma 4.1, then we obtain the conclusion. □

Proof of Theorem 4.2 For each $t \geqslant 0$ and any given $x(t) \in (-\infty, \infty)^n$, it is easy to check that there exist disjointed subsets $V_k(t)$, $k = 1, \cdots, m (1 \leqslant m \leqslant n)$ such that $V = \bigcup_{k=1}^{m} V_k(t)$ and each $\mathcal{G}_{V_k}(t) = \{V_k(t), \varepsilon_k(t)\}$ is either a complete graph or a connected but not complete graph. If $\mathcal{G}_V(0)$ is a complete graph, by Lemma 4.1, the conclusion holds. Otherwise, at each moment $t \geqslant 0$ consider the protocol (4.6) for all subsets $V_k(t)$, $1 \leqslant k \leqslant m$ with connected but not complete graphs.

If $\mathcal{G}_k(t)$ is a connected but complete graph, following the same argument below (4.6), we know that $d_{V_c(t)}(t)$ reduces by at least $2a$ after each step under the protocol (4.6). If $\mathcal{G}_k(t)$ is a complete graph, by Lemma 4.1, we know that $d_{V_c(t)}(t) \leqslant \varepsilon$ before it meets another subgroup. Since $d_{V_k(t)}(t) \leqslant |V_k(t)| \varepsilon \leqslant n\varepsilon$ when $\mathcal{G}_k(t)$ is a connected but not complete graph, we can get that under protocol (4.6), $\mathcal{G}_k(t)$ will become a complete graph after no more than $\left\lceil \frac{(n-1)\varepsilon}{2a} \right\rceil$ steps. Considering that during this period two subgroups may meet and become a new connected but not a complete graph, we know that under the protocols (4.6), all subgroups will become complete graphs after no more than $\left\lceil \frac{(n-1)^2\varepsilon}{2a} \right\rceil$ steps. By independence of $\xi_i(t)$, $i \in V$, $t > 0$, we know that

$$\mathbb{P}\left\{\text{protocol (4.6) occurs} \left\lceil \frac{(n-1)^2\varepsilon}{2a} \right\rceil \text{times}\right\} \geqslant p^{\lceil \frac{n(n-1)^2\varepsilon}{2a} \rceil} > 0$$

implying for any given $x(0) \in (-\infty, \infty)^n$, there exists a constant $L \leqslant \left\lceil \frac{(n-1)^2\varepsilon}{2a} \right\rceil$ such that

$$\mathbb{P}\{G_V(L) \text{ consists of complete graphs}\} \geqslant p^{\lceil \frac{n(n-1)^2\varepsilon}{2a} \rceil} > 0 \tag{4.7}$$

Denote $C(L) = \{\omega: G_V(L) \text{ consists of complete graphs}\}$, then by Lemma 4.2, there exists a finite time $\overline{T}_1$ which is $\sigma(\xi(1), \cdots)$ - measurable, and a constant $0 < p_0 < 1$ such that

$$\mathbb{P}\{d_V(L + \overline{T}_1) \leqslant \varepsilon\} = \mathbb{P}\{d_V(\overline{T}_1) \leqslant \varepsilon \mid C(L)\} \cdot \mathbb{P}C(L)\} \geqslant p_0 p^{\lceil \frac{n(n-1)^2\varepsilon}{2a} \rceil} > 0$$

and hence

$$\mathbb{P}\{d_V(L + \overline{T}_1) > \varepsilon\} \leqslant 1 - p_0 p^{\lceil \frac{n(n-1)^2\varepsilon}{2a} \rceil} < 1 \tag{4.8}$$

For a finite time T, define $U(T) = \{\omega: d_V(L+T) > \varepsilon\}$, $U = \{\omega: (4.1) \sim (4.2)$ does not reach quasi-synchronization in finite time$\}$. By (4.8),

$$\mathbb{P}\{U(\overline{T}_1)\} \leqslant 1 - p_0 p^{\lceil \frac{n(n-1)^2\varepsilon}{2a} \rceil} < 1$$

Since $x(0)$ is arbitrarily given in $(-\infty, \infty)^n$, considering the independence of $\sigma(\overline{T}_1)$ and $\sigma(\xi(\overline{T}_1 + 1), \cdots)$ and following the procedure of (4.8), we know there exists a finite time sequence $\overline{T}_1 \leqslant \overline{T}_2 \leqslant \cdots < \infty$ such that

$$\mathbb{P}\{U(\overline{T}_{m+1}) \mid U(\overline{T}_m)\} \leqslant 1 - p_0 p^{\lceil \frac{n(n-1)^2\varepsilon}{2a} \rceil},\ m \geqslant 1$$

Notice by Lemma 4.1 that once there is a finite time T that $d_V(T) \leqslant \varepsilon$, it will hold $d_V \leqslant \delta_2 - \delta_1 \leqslant \varepsilon$ thus $U(\overline{T}_{j+1}) \subset U(\overline{T}_j)$, $j \geqslant 1$ and hence

$$\begin{aligned}
\mathbb{P}\{U\} &\leqslant \mathbb{P}\left\{\bigcap_{m=1}^{\infty} U(\overline{T}_m)\right\} = \lim_{m\to\infty} \mathbb{P}\left\{\bigcap_{j=1}^{m} U(\overline{T}_j)\right\} \\
&= \lim_{m\to\infty} \prod_{j=1}^{m-1} \mathbb{P}\left\{U(\overline{T}_{j+1}) \,\Big|\, \bigcap_{l \leqslant j} U(\overline{T}_l)\right\} \cdot \mathbb{P}\{U(\overline{T}_1)\} \\
&= \lim_{m\to\infty} \prod_{j=1}^{m-1} \mathbb{P}\{U(\overline{T}_{j+1}) \mid U(\overline{T}_j)\} \cdot \mathbb{P}\{U(\overline{T}_1)\} \\
&\leqslant \lim_{m\to\infty} \left(1 - p_0 p^{\lceil \frac{n(n-1)^2\varepsilon}{2a} \rceil}\right)^m = 0
\end{aligned}$$

here the first equation holds since $\left\{\bigcap_{j=1}^{m} U(\overline{T}_j),\ m\geqslant 1\right\}$ is a decreasing sequence and $\mathbb{P}$ is a probability measure. As a result

$$\mathbb{P}\{(4.1)\sim(4.2)\text{ reach quasi-synchronization in finite time}\}=1-\mathbb{P}\{U\}=1$$

This completes the proof. □

Next, we will present the necessary part of the noise-induced synchronization, which shows that when the noise amplitude has a positive probability of exceeding ε, the system a. s. cannot reach quasi-synchronization.

Theorem 4.3 (Necessary condition for quasi-synchronization of the HK model with independent noise) Let $x(0)\in(-\infty,\infty)^n$, $\varepsilon>0$ are arbitrarily given. Suppose the noises $\{\xi_i(t),\ i\in V,\ t\geqslant 1\}$ are independent and there exist constants $\delta_2-\delta_1\geqslant\varepsilon$ and $0<q\leqslant 1$ such that $\mathbb{P}\{\xi_i(t)<\delta_1\}\geqslant q$ and $\mathbb{P}\{\xi_i(t)>\delta_2\}\geqslant q$, then a. s. the system (4.1) ~ (4.2) cannot reach quasi-synchronization.

Proof We only need to prove that, for any $T_0\geqslant 0$, there exists $t>T_0$ a. s. such that $d_V(t)>\varepsilon$ a. s. , i. e.

$$\mathbb{P}\left\{\bigcup_{T_0=0}^{\infty}\{d_V(t)\leqslant\varepsilon,\ t>T_0\}\right\}=0$$

Given any $T_0\geqslant 0$, by independence of $\xi_i(t)$, $i\in V$, $t\geqslant 1$, we have

$$\begin{aligned}&\mathbb{P}\{d_V(T_0+1)>\varepsilon\mid d_V(T_0)\leqslant\varepsilon\}\\&\geqslant\mathbb{P}\left\{\max_{x_i(T_0),i\in V}\xi_i(T_0+1)>\delta_2,\ \min_{x_i(T_0),i\in V}\xi_i(T_0+1)<\delta_1\right\}\\&\geqslant q^2\end{aligned}$$

Hence, $\mathbb{P}\{d_V(T_0+1)\leqslant\varepsilon\mid d_V(T_0)\leqslant\varepsilon\}\leqslant 1-q^2<1$, Similarly, for all $t>T_0$

$$\mathbb{P}\left\{d_V(t)\leqslant\varepsilon\,\Big|\,\bigcap_{T_0<l<t}\{d_V(l)\leqslant\varepsilon\}\right\}\leqslant 1-q^2$$

Noting $\mathbb{P}\{d_V(T_0)\leqslant\varepsilon\}\leqslant 1$, it has

$$\begin{aligned}\mathbb{P}\{d_V(t)\leqslant\varepsilon,\ t>T_0\}&=\mathbb{P}\left\{\bigcap_{t=T_0+1}^{\infty}\{d_V(t)\leqslant\varepsilon\}\right\}\\&=\lim_{m\to\infty}\mathbb{P}\left\{\prod_{t=T_0+1}^{m}\{d_V(t)\leqslant\varepsilon\}\right\}\\&\leqslant\lim_{m\to\infty}\prod_{t=T_0+1}^{m}\mathbb{P}\left\{d_V(t)\leqslant\varepsilon\,\Big|\,\bigcap_{T_0\leqslant l<t}\{d_V(l)\leqslant\varepsilon\}\right\}\\&\leqslant\lim_{m\to\infty}(1-q^2)^m=0\end{aligned}$$

This completes the proof. □

Theorem 4.3 shows that when the noise amplitude has a positive probability to exceed the confidence threshold, the cluster will be destroyed by noise. Thus, the essence of noise-induced synchronization is that noise can drive the system towards a synchronized state, and the system is capable of maintaining that state. When the noise is large and exceeds a critical amplitude, the system is fluctuating severely such that the synchronized state cannot be maintained anymore.

4.3 Simulations

In this part, we will present some simulation results to verify the main theoretical results in this chapter. First, we present a fragmentation of noise-free HK model. We take $n = 20$, $\varepsilon = 5$ and the initial states are randomly generated on $[0, 50]$. Fig. 4.1 shows the formation of four clusters. We then add independent noises which are uniformly distributed on $[\delta_1, \delta_2]$ to the agents. By Theorem 4.2, when $0 < \delta_2 - \delta_1 \leqslant \varepsilon$, the system will almost surely achieve quasi-synchronization. Let $\delta_1 = -2$, $\delta_2 = 2.1$, then Fig. 4.2 clearly displays the quasi-synchronization picture. Next, we consider the case when the noise amplitude $\delta_2 - \delta_1$ exceeds the critical value ε. For a better demonstration, we simply show a synchronized system will divide in the presence of larger noise. After taking $n = 10$, $x_i(0) = 0$, $1 \leqslant i \leqslant 10$ and $\varepsilon = 5$. Let $\delta_1 = -3, \delta_2 = 3.5$, our Fig. 4.3 shows the clear separation of the system.

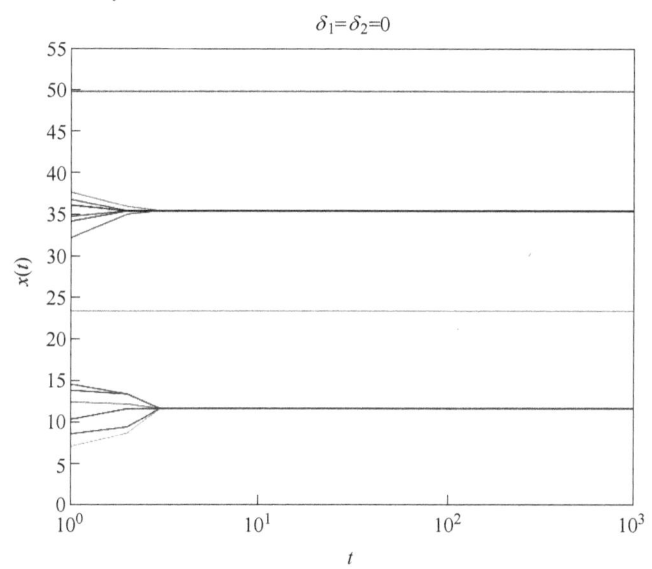

Fig. 4.1 Evolution of system (4.1) ~ (4.2) of 20 agents without noise

(The initial system states are randomly generated on $[0, 50]$, confidence threshold $\varepsilon = 5$)

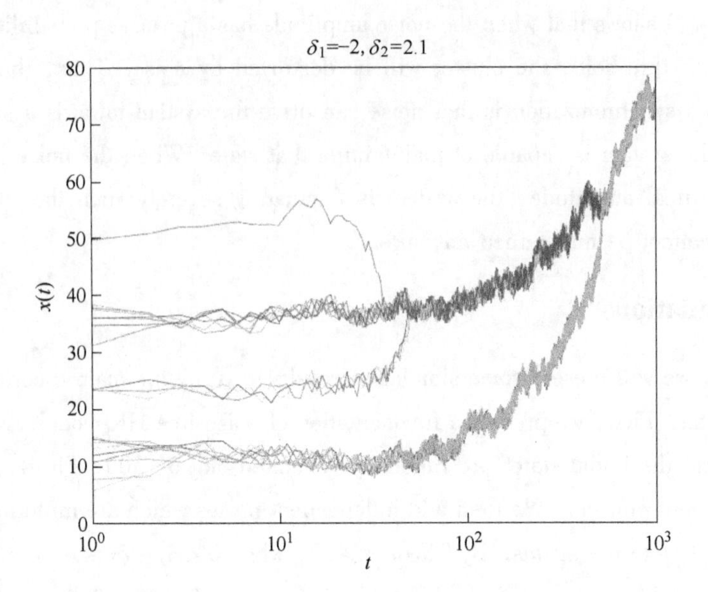

Fig. 4. 2　Evolution of system (4. 1) ~ (4. 2) of 20 agents with noise uniformly distributed on [−2, 2. 1]

(The initial conditions are identical with those in Fig. 4. 1, except that adding noises are uniformly distributed on [−2, 2. 1])

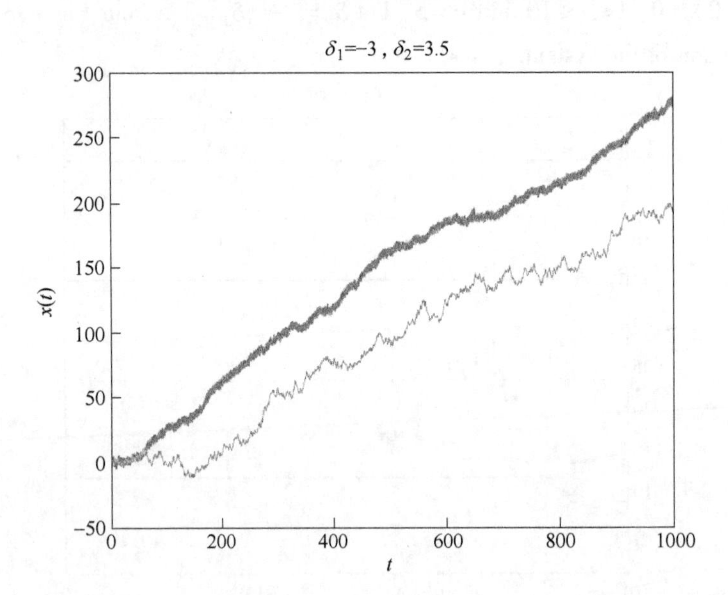

Fig. 4. 3　Evolution of system (4. 1) ~ (4. 2) of 10 agents with noise uniformly distributed on [−3, 3. 5]

(The initial system states are identically taken to be 0; the confidence threshold $\varepsilon = 5$)

4.4 Notes

In this chapter, we mainly established a rigorous theoretical analysis for noise-induced synchronization of the HK model in the full state-space. By investigating the graph property of the HK dynamics, we completely resolved this problem. Moreover, a critical noise amplitude for the induced synchronization is obtained. The analysis skill that we developed for the graph property of the HK model will provide further tools for studying synchronization problems in noisy HK-based dynamics. Moreover, given the flexible generalizability of our results, we hope our analysis will stimulate much further research on noise-induced synchronization phenomena in physical, biological, and social self-organizing systems.

References

[1] Etesami S, Basar T. Game-Theoretic Analysis of the Hegselmann-Krause Model for Opinion Dynamics in Finite Dimensions[J]. IEEE Transactions on Automatic Control, 2015, 60 (7): 1886 ~ 1897.

[2] Hegselmann R, Krause U. Opinion dynamics and bounded confidence models, analysis, and simulation[J]. Journal of Artificial Societies and Social Simulation, 2002, 5 (3): 1 ~ 33.

[3] Lorenz J. A stabilization theorem for continuous opinion dynamics[C]. Physica A: Statistical Mechanics and its Applications, 2005, 355 (1): 217 ~ 223.

[4] Krause U. A discrete nonlinear and non-automonous model of consensus formation. Communications in Difference Equations[J]. Amsterdam: Gordon and Breach Publisher, 2000, 227 ~ 238.

[5] Su W, Chen G, Hong Y. Noise leads to quasi-consensus of Hegselmann-Krause opinion dynamics [J]. Automatica, 2017, 85: 448 ~ 454.

[6] Pineda M, Toral R, Hernandez-Garcia E. The noisy Hegselmann-Krause model for opinion dynamics[J]. The European Physical Journal B, 2013, 86 (12): 1 ~ 10.

[7] Su W, Yu Y. Free information flow benefits truth seeking[J]. Journal of Systems Science and Complexity, 2018, 31 (4): 964 ~ 974.

[8] Wang C, Li Q, E W, Chazelle B. Noisy Hegselmann-Krause Systems: Phase Transition and the 2R-Conjecture[C]. 2016 IEEE 55th Conference on Decision and Control, 2016, 2632 ~ 2637.

[9] Garnier J, Papanicolaou G, Yang T. Consensus convergence with stochastic effects[J]. Vietnam Journal of Mathematics, 2017, 45: 51 ~ 75.

5 Heterogeneous HK Model with Environment and Communication Noise

Because the inter-agent topology of the HK model is time-varying and deter-mined by the agents' states, whereas the agents' states depend on the topology, the theoretical analysis of the HK model is difficult. The current analysis of the HK model focuses on the most basic property of convergence. For the homogeneous HK model-where all agents have the same confidence bound-the convergence and convergence rate have been well studied[1~5]. Also, there exists some theoretical research on varieties of the homogeneous HK model, such as the systems with decaying confidence[6], with distance-dependent interaction weight[7], or with continuous agents[8]. For the heterogeneous HK model where the confidence bounds of the agents can be different-Su et al. [9] prove that partial agents in the system will reach static states after a finite time, however cannot prove the convergence of other agents. Besides, the opinions of all agents are shown to be convergent if each agent maintains communication with others during a long enough period of time[10], or if the confidence bound of each agent is either 0 or 1[11]. However, the convergence of the general heterogeneous HK model without additional conditions is still an open problem (Conjecture 2.1 in [12]), despite it having been supported by simulations[13]. The analysis of the heterogeneous HK model is particularly important since there are always differences between individuals, contributing to one motivation of this chapter.

Another motivation of this chapter is to study the collective behavior of the HK model affected by noise. There is a consensus that all natural systems are inextricably affected by noise[14]. Actually, how noise affects the collective behavior of a complex system has garnered considerable interest from researchers and developers in differing fields. Generally, the noise in engineering systems may break their ordered structures, in which case one wishes to reduce the effect of the noise. However, in many natural and social systems, the noise may drive the systems to produce ordered structure[14, 15]. As a matter of fact, the actual opinions of individuals are inevitably influenced by the randomness during opinion transmission and evolution, which could be attributable to the many exogenous factors like T. V. , blogs, and newspapers, or the communication between individuals. Thus, it has been recognized by several studies that randomness is an essential factor for the investigation of opinion dynamics in reality[16~22]. In many studies, an interesting phenomenon was found where the noise in some situations could play a positive role in enhancing the synchronization or reducing the disagreement of opinions. Yet almost all of these findings were based on simulations, and the theoretical analysis is limited. In Chapters 3 and 4, homogeneous HK models with environment noise was studied, but the analysis method cannot be applied to the heterogeneous case. This chapter will analyze the heterogeneous HK model with environment noise by means of a completely different method. Also, to be more practical, this chapter considers different types of heterogeneous HK models that may be affected by communication noise and global information. The communication noise is caused by individuals potentially not expressing their own opinion or not accurately understanding the opinions of others, while the global information denotes the background opinion modeled by the average opinion of all individuals.

5.1 Models and Definitions

The original HK model assumes that there are $n\,(n\geqslant 3)$ individuals or agents in a group. Each individual i has a time-varying opinion $x_i(t)\in[0,1]$, and can only communicate opinions with his/her friends, which is defined by where the difference of opinions is not bigger than a confidence threshold $r_i\in(0,1]$. This mechanism is based on a practical phenomenon where one individual is not willing to accept the opinion of another if he/she feels their opinions have a large gap. Let

$$N_i(t)=\{1\leqslant j\leqslant n \mid |x_j(t)-x_i(t)|\leqslant r_i\}$$

denote the neighbor set of agent i at time t. Here we note that an individual's neighbor set contains himself/herself. The evolution of opinions of the HK model accords to the following dynamics:

$$x_i(t+1)=\frac{1}{|N_i(t)|}\sum_{j\in N_i(t)} x_j(t),\ i=1,\cdots,n,\ t\geqslant 0 \tag{5.1}$$

Where $|S|$ denoting the cardinal number of a set S.

If $r_1=\cdots=r_n$, then the system (5.1) is called the homogeneous HK model, otherwise it is referred to as the heterogeneous HK model. The HK model is a typical self-organized system that has attracted a significant amount of interest, but has shown difficult to analyze. Currently, the analysis of the HK model focuses on the homogeneous case, while the analysis of the heterogeneous case is almost lacking.

5.1.1 Heterogeneous Models with Environment Noise

The dynamics of opinions in real societies is also affected by many exogenous factors such as T. V. , blogs, newspapers, and so on[24]. Some HK-type systems under the effect of exogenous factors have been considered in recent years[24-26]. This chapter considers exogenous factors as environment noises. Following previous work in noisy opinion dynamics[16, 20, 21], we confine the values of an individual's opinion to the interval $[0,1]$. Let $\prod_{[0,1]}(\cdot)$ denote the projection onto the interval $[0,1]$, i. e. , for any $x \in R$,

$$\prod_{[0,1]}(x) = \begin{cases} 1 & \text{if } x > 1 \\ x & \text{if } x \in [0,1] \\ 0 & \text{otherwise} \end{cases}$$

Consider a basic HK model with environment noise as follows: Denote $V = \{1,2,\cdots,n\}$ as the set of all agents with $n \geqslant 3$. For all $i \in V$ and $t \geqslant 0$, let

$$x_i(t+1) = \prod_{[0,1]}\left(\frac{1}{|N_i(t)|}\sum_{j\in N_i(t)} x_j(t) + \xi_i(t)\right) \tag{5.2}$$

where $\xi_i(t) \in [-\eta,\eta]$ is a bounded noise with $\eta > 0$ being a constant.

A natural consideration is that all agents may be affected in reality by the background opinion. Accordingly, an interesting problem is how the background opinion affects the collective behavior of the opinions of agents. For simplicity, this chapter models the background opinion as the average opinion $x_{\text{ave}}(t) = \frac{1}{n}\sum_{i=1}^{n} x_i(t)$, and each agent i has a belief factor $\omega_i \in (0,1)$ in the global information $x_{\text{ave}}(t)$. The HK model with global information and environment noise is formulated as

$$x_i(t+1) = \prod_{[0,1]}\left(\omega_i x_{\text{ave}}(t) + \frac{1-\omega_i}{|N_i(t)|}\sum_{j\in N_i(t)} x_j(t) + \xi_i(t)\right)$$

Let $x(t) \triangleq (x_1(t),\cdots,x_n(t))$. To be more practical we consider a wide class of noises which contain not only independent noises but also correlated noises. For systems (5.2) and (5.3), let $\Omega^t = \Omega_n^t \subseteq R^{n\times(t+1)}$ be the sample space of $(\xi_i(t'))_{0\leqslant t'\leqslant t, i\in V}$, and $\mathcal{F}^t = \mathcal{F}_n^t$ be its Borel σ-algebra. Additionally, we define Ω^{-1} be the empty set. Let $\mathbb{P} = \mathbb{P}_n$ be the probability measure on $\mathcal{F}^\infty$ for $(\xi_i(t'))_{t'\geqslant 0, i\in V}$, so the probability space is written as $(\Omega^\infty, \mathcal{F}^\infty, \mathbb{P})$. We assume that the noises $\{\xi_i(t)\}$ satisfy the following assumption.

(A1) For any $t\geqslant 0$ and any states $x(0),\cdots,x(t)\in[0,1]^n$, the joint probability density of $(\xi_1(t),\cdots,\xi_n(t))$ has a positive lower bound, i. e. , there exists a constant $\underline{\rho}>0$ such that

$$P\left(\bigcap_{i=1}^{n}\{\xi_i(t)\in[a_i,b_i]\}\mid\forall x(0),\cdots,x(t)\in[0,1]^n\right)\geqslant\underline{\rho}\prod_{i=1}^{n}(b_i-a_i)$$

for any $t\geqslant 0$ and real numbers a_i and b_i satisfying $-\eta\leqslant a_i<b_i\leqslant\eta$.

The positive lower bound in (A1) simply means that for any $t\geqslant 0$, all individuals are affected by noise, and the noise has a positive probability density over $[-\eta,\eta]$. It is easy to see that if $\xi_i(t)$ is uniformly and independently distributed in $[-\eta,\eta]$, then it satisfies (A1). Some other bounded noises also satisfy (A1), such as the truncated Gaussian noise[27], as well as the discrete time version of frequency fluctuations generated by sinusoidal functions and Wiener processes[28].

5.1.2 Heterogeneous Models with Communication Noise

In reality, communication between individuals may be subject to noise because individuals may not express their own opinion or accurately understand the opinions of others. The heterogeneous HK dynamics with communication noise can be formulated as

$$x_i(t+1)=\prod_{[0,1]}\left(\frac{1}{|N_i(t)|}\sum_{j\in N_i(t)}[x_j(t)+\zeta_{ji}(t)]\right)\tag{5.3}$$

for all $i\in V$ and $t\geqslant 0$, where denotes the communication noise from agent j to i at time t, with $\zeta_{ji}(t)\equiv 0$.

Similar to (5.3) we also consider an HK model with global information and communication noise as follows: for $i\in V$, $t\geqslant 0$,

$$x_i(t+1)=\prod_{[0,1]}\left(\omega_i x_{\text{ave}}(t)+(1-\omega_i)\frac{1}{|N_i(t)|}\sum_{j\in N_i(t)}[x_j(t)+\zeta_{ji}(t)]\right)$$

For the systems (5.3) and (5.4), let $\Omega^t=\Omega_n^t$ be the sample space of $(\zeta_{ji}(t'))$ where $0\leqslant t'\leqslant t$, $i\in V$, $j\in N_i(t')$, and $\mathcal{F}^t=\mathcal{F}_n^t$ be its Borel σ-algebra. Additionally, we define Ω^{-1} as the empty set. Let $\mathbb{P}=\mathbb{P}_n$ be the probability measure on $\mathcal{F}^\infty$ for $(\zeta_{ji}(t'))_{t'\geqslant 0,\, i\in V,\, j\in N_i(t')}$, so the probability space is written as $(\Omega^\infty,\mathcal{F}^\infty,\mathbb{P})$ Define the set of agent pairs $\varepsilon(t)$ by $\varepsilon(t)\triangleq\{(i,j):i\in V,j\in N_i(t)\setminus\{i\}\}$ Similar to (A1), we assume the noises $\{\zeta_{ji}(t)\}$ satisfying the following assumption.

(A2) For any $t\geqslant 0$ and any states $x(0),\cdots,x(t)\in[0,1]^n$, if $\varepsilon(t)$ is not empty, then, the joint probability density of $\{\zeta_{ji}(t)\}_{(i,j)\in\mathcal{E}(t)}$ has a positive lower bound, i. e., there exists a constant $\underline{\rho}>0$ such that

$$P\Big(\bigcap_{(i,j)\in\mathcal{E}(t)}\{\zeta_{ji}(t)\in[a_i^j,b_i^j]\}\mid\forall x(0),\cdots,x(t)\Big)\geqslant\underline{\rho}\prod_{(i,j)\in\mathcal{E}(t)}(b_i^j-a_i^j)$$

for any $t\geqslant 0$ and real numbers a_i^j and b_i^j satisfying $-\eta\leqslant a_i^j<b_i^j\leqslant\eta$.

5.1.3 Definitions

Define

$$d_V(t)=\max_{i,j\in V}|x_i(t)-x_j(t)|$$

to be the maximum opinion difference at time t. Let

$$\underline{d}_V=\liminf_{t\to\infty}d_V(t)\text{ and }\overline{d}_V=\limsup_{t\to\infty}d_V(t)$$

denote the lower limit and upper limit of the maximum opinion difference respectively.

Similar to Chapters 3 and 4, we relax the concept of synchronization to *quasi-synchronization* which is defined as follows:

Definition 5.1 We say that quasi-synchronization is asymptotically reached if $\overline{d}_V\leqslant\min_{1\leqslant i\leqslant n}r_i$.

From Definition 5.1, when quasi-synchronization is reached, any two agents can communicate directly in the limit state. In fact, in Theorems 5.1, 5.2, 5.3, and 5.5 below, wherever a bound $\overline{d}_V\leqslant\min r_i$ exists, almost surely there is some finite time t_0 such that $d_V\leqslant\min r_i$ for all $t\geqslant t_0$. In other words, almost surely agents communicate directly from some time onwards.

5.2 Main Results for Quasi-Synchronization

In this section we will first analyze systems (5.2) and (5.3) which exhibit a phase transition for quasi-synchronization behavior depending on the noise amplitude η. Let $r_{\min}=\min_{1\leqslant i\leqslant n}r_i$ and $r_{\max}=\max_{1\leqslant i\leqslant n}r_i$. Also, for any $\alpha>0$ we set

$$c_\alpha^1=c_\alpha^1(\eta):=\begin{cases}1 & \text{if }\eta>\max\left\{\dfrac{r_{\min}}{2},\dfrac{r_{\max}}{n}\right\}\\ 2\eta-\alpha & \text{otherwise}\end{cases}$$

With the above definitions we first give our main result for system (5.2) as follows:

Theorem 5.1 (Phase transition and switching interval in the heterogeneous HK model with environment noise) Consider system (5.2) satisfying (A1) and $r_{\min} < 1$. Then for any initial state $x(0) \in [0,1]^n$, almost surely (a. s.)

$$\bar{d}_V \begin{cases} = 2\eta, & \text{if } \eta \leqslant \frac{r_{\min}}{2} \\ \geqslant \min\{2\eta, 1\}, & \text{if } \eta > \frac{r_{\min}}{2} \\ = 1, & \text{if } \eta > \max\left\{\frac{r_{\min}}{2}, \frac{r_{\max}}{n}\right\} \end{cases} \tag{5.4}$$

Also, for any constant $\alpha > 0$, if we set $\tau_0 = 0$, and τ_k to be the stopping time as

$$\tau_k = \begin{cases} \min\{s > \tau_{k-1} : d_V(s) \leqslant \alpha\}, & \text{if } k \text{ is odd} \\ \min\{s > \tau_{k-1} : d_V(s) \geqslant c_\alpha^1\}, & \text{if } k \text{ is even} \end{cases} \tag{5.5}$$

then for all $k \geqslant 0$ and $t \geqslant 0$ we have

$$\mathbb{P}(\tau_{k+1} - \tau_k > t) \leqslant (1 - \lambda_1)^{\lfloor t/\Lambda_1 \rfloor} \tag{5.6}$$

where $\lambda_1 \in (0,1)$ and $\Lambda_1 > 0$ are constants only depending on n, η, ρ, α, and r_i, $i \in V$.

The inequality (5.6) denotes that $d_V(t)$ will switch between $(0, \alpha]$ and $[c_\alpha^1, 1]$ infinitely often, and the switching interval is a random variable depending on n, η, ρ, α, and r_i, $i \in V$. However, the specific dependencies are very non-linear, and difficult to describe even through simulations.

Remark 5.1 From (5.4), the upper limit of the maximum opinion difference $\bar{d}_V$ has a phase transition at the point $\eta = \frac{r_{\min}}{2}$, providing $r_{\max} \leqslant \frac{n r_{\min}}{2}$. This result implies that the maximum opinion difference depends on the minimal confidence threshold among all individuals. Thus, by comparing the homogeneous and heterogeneous cases of system (5.2) with the same average confidence bound, the heterogeneity of individuals is harmful to synchronization, which may be the reason why the synchronization of opinions is hard to reach in reality, and even within that of a small group.

We also provide some simulations for Theorem 5.1. Consider the system (5.2) with n agents whose initial opinions are all set to be 0.5. For the confidence bounds of the agents, we set $r_1 = r_{\min} = 0.05$ and $r_n = r_{\max} = 0.45$, and choose $\{r_i\}_{i=2}^{n-1}$ randomly and uniformly from $[0.05, 0.45]$. Suppose the noises $\{\xi_i(t)\}$ are independently and uniformly distributed in $[-\eta, \eta]$. All simulations run up to 10^6 steps. We first choose $n =$

20 and $\eta = 0.025 = \frac{r_{\min}}{2}$, and the value of $d_V(t)$ is shown in Fig. 5.1. In this figure, it can be observed that $\overline{d}_V = 0.05 = 2\eta = r_{\min}$, which is consistent with (5.4) and the system (5.2) reaches quasi-synchronization. Second we choose $n = 20$ and $\eta = 0.1 > \max\left\{\frac{r_{\min}}{2}, \frac{r_{\max}}{n}\right\}$, and the value of $d_V(t)$ is presented in Fig. 5.2. In this figure it can be seen that $\overline{d}_V = 1$, which is also consistent with (5.4). Finally we consider a small group with $n = 4$ and $\eta = 0.1 \in \left(\frac{r_{\min}}{2}, \frac{r_{\max}}{n}\right)$, and the value of $d_V(t)$ is provided in Fig. 5.3. Differing from Fig. 5.2, it seems that $\overline{d}_V < 1$, so the behavior of the system with a small number of agents is quite different from a large number of agents.

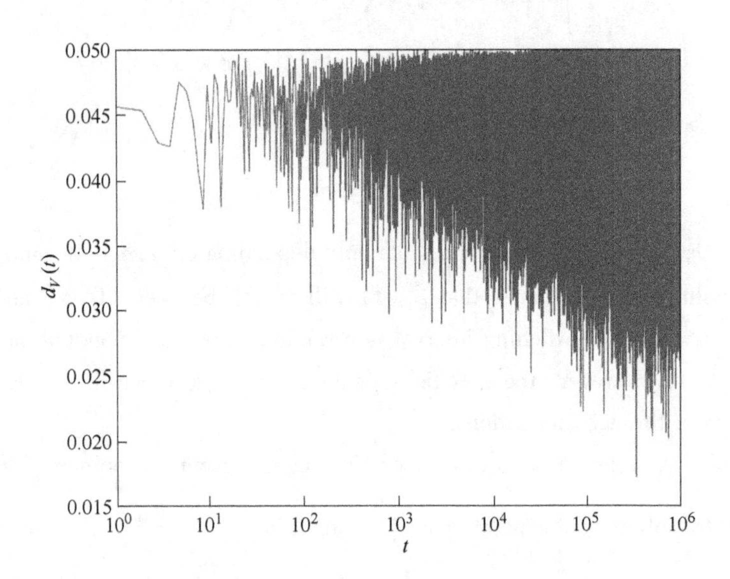

Fig. 5.1　The value of $d_V(t)$ under system (5.2) with $n = 20$ and $\eta = 0.025$

By Theorem 5.1 and Definition 5.1 we get the following corollary concerning the critical noise amplitude and convergence rate for quasi-synchronization:

Corollary 5.1 (Critical noise amplitude and convergence rate for quasi-synchronization of system (5.2))　Consider the system (5.2) satisfying (A1) and $r_{\min} < 1$. Then, for any initial state, the system asymptotically reaches quasi-synchronization a. s. if and only if $\eta \leqslant \frac{r_{\min}}{2}$.

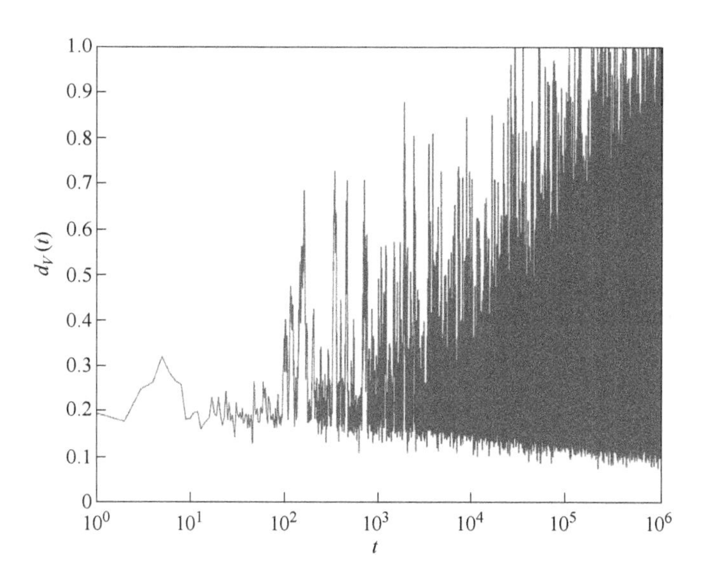

Fig. 5.2 The value of $d_V(t)$ under system (5.2) with $n = 20$ and $\eta = 0.1$

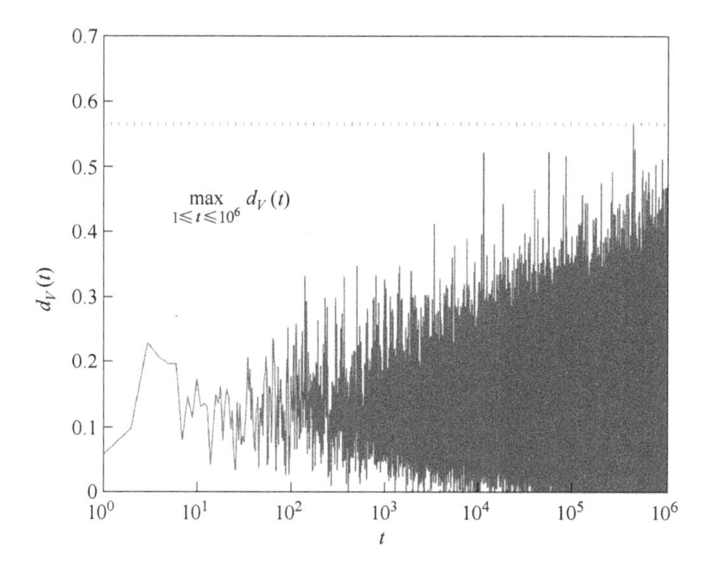

Fig. 5.3 The value of $d_V(t)$ under system (5.2) with $n = 4$ and $\eta = 0.1$

Moreover, if $\eta \leqslant \frac{r_{\min}}{2}$, let τ be the minimal t satisfying $d_V(t) \leqslant 2\eta$. Then, $d_V(t) \leqslant 2\eta$ for all $t \geqslant \tau$, and there exist constants $\lambda_2 \in (0,1)$ and $\Lambda_2 > 0$ depending on n, η, ρ, and r_i, $i \in V$ only such that

$$\mathbb{P}(\tau > s) \leqslant (1-\lambda_2)^{\lfloor s/\Lambda_2 \rfloor}, \ \forall s \geqslant 0$$

Next, we give our main result for system (5.3). For any $\eta > \frac{r_{\min}}{2}$, let $\underline{w}_\eta = \min\{w_i : r_i < 2\eta\}$ and

$$c_\eta^2 = \max\left\{2\eta, \ \min\left\{\frac{n\eta}{(n-1)\underline{\omega}_\eta}, \ n\eta\right\}\right\}$$

Also, for any $\alpha > 0$ we define

$$c_\alpha^3 \triangleq \begin{cases} 2\eta - \alpha, & \text{if } \eta \leqslant \frac{r_{\min}}{2} \\ \min\{c_\eta^2 - \alpha, \ 1\}, & \text{if } \eta > \frac{r_{\min}}{2} \end{cases}$$

Theorem 5.2 (Phase transition and switching interval for the heterogeneous HK model with environment noise and global information) Consider system (5.3) satisfying (A1) and $r_{\min} < 1$. Then for any initial state $x(0) \in [0,1]^n$, almost surely

$$\bar{d}_V \begin{cases} = 2\eta, & \text{if } \eta \leqslant \frac{r_{\min}}{2} \\ \geqslant \min\{c_\eta^2, \ 1\}, & \text{if } \eta > \frac{r_{\min}}{2} \end{cases} \tag{5.7}$$

Also, for any constant $\alpha > 0$, if we define $\tau_0 = 0$, and $\{\tau_k\}_{k \geqslant 1}$ as same as (5.5) but using c_α^3 instead of c_α^1, then for all $k \geqslant 0$ and $t \geqslant 0$ we have

$$\mathbb{P}(\tau_{k+1} - \tau_k > t) \leqslant (1-\lambda_3)^{\lfloor t/\Lambda_3 \rfloor} \tag{5.8}$$

where $\lambda_3 \in (0,1)$ and $\Lambda_3 > 0$ are constants only depending on n, η, $\underline{\rho}$, α and ω_i, r_i, $i \in V$.

Remark 5.2 Differing from system (5.2), the system (5.3) has a global aver-age opinion $x_{\text{ave}}(t)$, which means all agents remain connected for all time. In Theorem 5.2, if we let $\underline{w}_\eta$ tend to zero, then $\bar{d}_V \geqslant \min\{n\eta, 1\}$ for the case when $\eta > \frac{r_{\min}}{2}$. Thus, the results of Theorems 5.1 and 5.2 are consistent for the case when $\eta \geqslant \max\left\{\frac{r_{\min}}{2}, \ \frac{1}{n}\right\}$. However, Theorem 5.1 is not a special case of Theorem 5.2 because the former cannot be included by the latter for the case when $\max\left\{\frac{r_{\min}}{2}, \frac{r_{\max}}{n}\right\} < \eta < \max\left\{\frac{r_{\min}}{2}, \frac{1}{n}\right\}$.

Similar to Theorem 5.1 we also provide some simulations for Theorem 5.2. Consider the system (5.3) with $n = 20$ agents whose initial opinions, confidence bounds, and noises have the same configurations as Fig. 5.1 and Fig. 5.2. Assume $\omega_1 = \cdots = \omega_n = 0.1$. All simulations run up to 10^6 steps. We first choose $\eta = 0.025$ as the same as Fig. 5.1, and the value of $d_V(t)$ is shown in Fig. 5.4. This figure displays a similar behavior of $d_V(t)$ as Fig. 5.1. Second, we choose $\eta = 0.1$ as the same as Fig. 5.2, and the value of $d_V(t)$ is shown in Fig. 5.5. Comparing this figure to Fig. 5.2, the system (5.3) has a smaller $\overline{d}_V$ than the system (5.2), signifying that the global average opinion can reduce the difference of opinions.

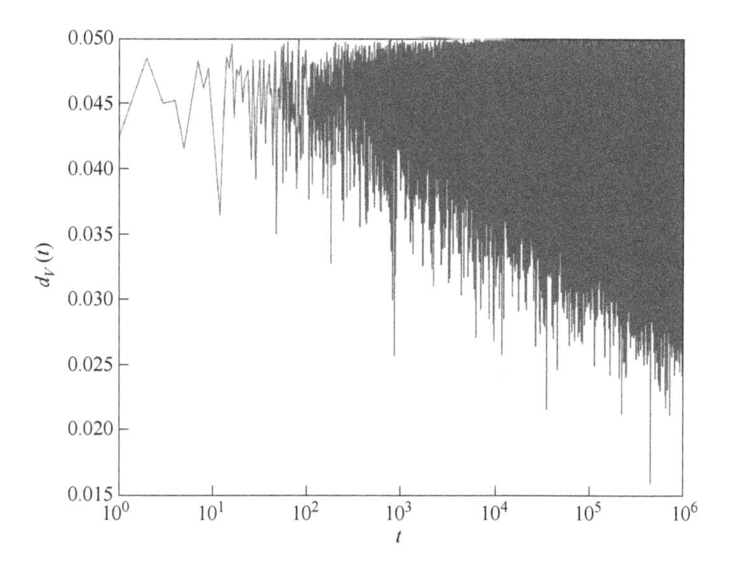

Fig. 5.4 The value of $d_V(t)$ under system (5.3) with $\eta = 0.025$

Similar to Corollary 5.1 we can get the following corollary concerning the critical noise amplitude and convergence rate for quasi-synchronization of system (5.3):

Corollary 5.2 (Critical noise amplitude and convergence rate for quasi-synchronization of system (5.3)) Consider the system (5.3) satisfying (A1) and $r_{\min} < 1$. Then, for any initial state, the system asymptotically reaches quasi-synchronization a. s.

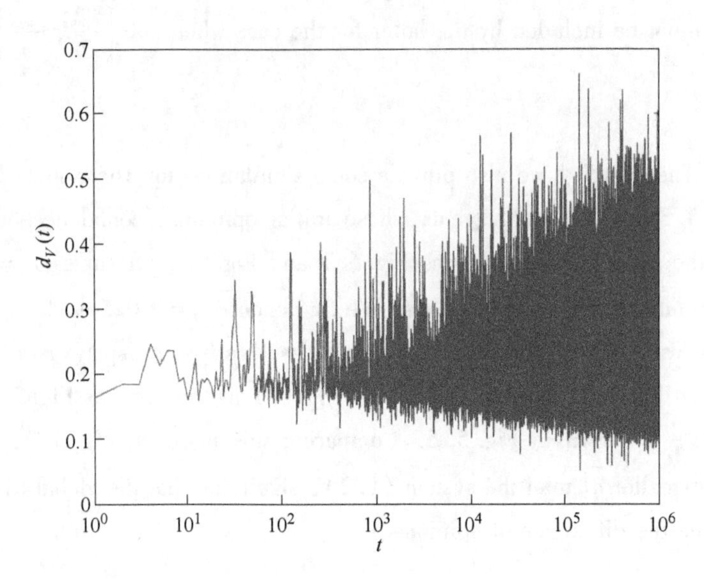

Fig. 5.5 The value of $d_V(t)$ under system (5.3) with $\eta = 0.1$

if and only if $\eta \leqslant \frac{r_{\min}}{2}$.

Moreover, if $\eta \leqslant \frac{r_{\min}}{2}$, let τ be the minimal t satisfying $d_V(t) \leqslant 2\eta$. Then, $d_V(t) \leqslant 2\eta$ for all $t \geqslant \tau$, and there exist constants $\lambda_4 \in (0,1)$ and $\Lambda_4 \in (0,1)$ depending on n, η, $\underline{\rho}$ and ω_i, r_i, $i \in V$ only such that

$$\mathbb{P}(\tau > s) \leqslant (1 - \lambda_4)^{\lfloor s/\Lambda_4 \rfloor}, \quad \forall s \geqslant 0$$

Remark 5.3 Combining Corollary 5.2 with Corollary 5.1 shows that the background opinion or global information does not affect the critical noise amplitude of quasi-synchronization.

Next, we consider the HK model with global information and communication noise which is much more complex than the HK model with global information and environment noise, since some agents may be not affected by noise if they have no neighbor except themselves. For $1 \leqslant i \neq j \leqslant n$, define $a_i \triangleq \frac{(n-1)(1-\omega_i)\eta}{n}$, $a_{ij} \triangleq a_i + a_j = \frac{(n-1)(2-\omega_i-\omega_j)\eta}{n}$,

$$h_{ij} \triangleq \begin{cases} \frac{(n-1)\eta}{n}\left[(1-\omega_i)^2 + \frac{\omega_i(\omega_j-\omega_i)}{n}\right], & \text{if } r_i < a_i \\ (1-\omega_i)\eta\left[\frac{1-\omega_i}{n} + \frac{n-2}{n-1}\right] + \frac{\omega_i(n-1)(\omega_j-\omega_i)\eta}{n^2}, & \text{if } r_i \in [a_i, a_{ij}) \\ \frac{(n-1)\eta}{n}\left[1-\omega_i + \frac{\omega_j-\omega_i}{n}\right], & \text{if } r_i \geqslant a_{ij} \end{cases}$$

and

$$c_{ij} = \min\{h_{ij} + h_{ji},\ 1-(a_i-h_{ij})I_{\{a_i>h_{ij}\}} - (a_j-h_{ji})I_{\{a_j>h_{ji}\}}\}$$

where $I_{\{\cdot\}}$ denotes the indicator function.

Theorem 5.3 (Analytical results for the heterogeneous HK model with communication noise and global information) Consider the system (5.4) satisfying (A2). Then for any initial state, almost surely

$$\bar{d}_V \begin{cases} = \max\limits_{i\neq j} a_{ij}, & \text{if } \eta \leqslant \min\limits_{i\neq j} \frac{nr_i}{(n-1)(2-\omega_i-\omega_j)} \\ \geqslant \min\{\max\limits_{i\neq j}\{a_{ij}, c_{ij}\}, 1\}, & \text{otherwise} \end{cases}$$

The result in Theorem 5.3 seems very complex. If we consider the large population that $n\to\infty$, and all agents has a same belief factor ω^* in the average opinion $x_{\text{ave}}(t)$, then by Theorem 5.3 we can get

$$\bar{d}_V \begin{cases} = 2(1-\omega^*)\eta, & \text{if } \eta \leqslant \frac{r_{\min}}{2(1-\omega^*)} \\ \geqslant \min\{2(1-\omega^*)\eta,\ 1\}, & \text{otherwise} \end{cases}$$

In this case, we can see that $\bar{d}_V$ still depends on the minimal confidence bound $r_{\min}$, though the phase transition is unknown. Also, it is shown that if $\eta \leqslant \frac{r_{\min}}{2}$, then $\bar{d}_V = 2\eta(1-\omega^*)$ is dependent on ω^*, whereas $\bar{d}_V = 2\eta$ in Theorem 5.2 is independent of $\{\omega_i\}$. The reason for this difference is that the noise in system (5.4) has a product with $1-\omega_i$, whereas the noise in system (5.3) does not have a product with $1-\omega_i$.

All the above results study the upper limit $\bar{d}_V$ of the maximum opinion difference. In fact, for the lower limit $\underline{d}_V$ we can get the following simple result:

Theorem 5.4 (The lower limit of maximum opinion difference in heterogeneous noisy HK models) Consider systems (5.2) and (5.3) satisfying (A1), and system (5.4) satisfying (A2). Then for any initial state, $\underline{d}_V = 0$ a. s..

Theorem 5.4 indicates that the opinions of all agents can reach almost consensus at infinite moments, however the consensus is not a stable state and the opinions may diverge under the influence of noise.

5.2.1 Proofs of Theorems 5.1 and 5.2, and Corollaries 5.1 and 5.2

We adopt the method of "transforming the analysis of a stochastic system into the design of control algorithms" first proposed by[23]. This method requires the construction of some new systems to help with the analysis of the noisy HK models. For $i \in V$ and $t \geqslant 0$, let

$$x_i(t) = \begin{cases} \dfrac{1}{|N_i(t)|}\sum\limits_{j \in N_i(t)} x_j(t), & \text{for protocols (5.2) and (5.3)} \\ \omega_i x_{\text{ave}}(t) + \dfrac{1-\omega_i}{|N_i(t)|}\sum\limits_{j \in N_i(t)} x_j(t), & \text{for protocols (5.3) and (5.4)} \end{cases} \tag{5.9}$$

From this definition the systems (5.2) and (5.3) can be rewritten as

$$x_i(t+1) = \prod_{[0,1]}(x_i(t) + \xi_i(t))$$

To analyze systems (5.2) and (5.3), we construct two robust control systems as follows. For $i \in V$ and $t \geqslant 0$, let $\delta_i(t) \in (0, \eta)$ be an arbitrarily given real number, $u_i(t) \in [-\eta + \delta_i(t), \eta - \delta_i(t)]$ denotes a bounded control input, and $b_i(t) \in [-\delta_i(t), \delta_i(t)]$ denotes the parameter uncertainty. For protocol (5.2) we construct a control system that for all $i \in V$ and $t \geqslant 0$

$$x_i(t+1) = \prod_{[0,1]}\left(\frac{1}{N_i(t)}\sum_{j \in N_i(t)} x_j(t) + u_i(t) + b_i(t)\right)$$

Similarly, for protocol (5.3) we construct a control system that for all $i \in V$ and $t \geqslant 0$

$$x_i(t+1) = \prod_{[0,1]}\left(\omega_i x_{\text{ave}}(t) + \frac{1-\omega_i}{|N_i(t)|}\sum_{j \in N_i(t)} x_j(t) + u_i(t) + b_i(t)\right)$$

By (5.9), the systems (5.10) and (5.10) can be rewritten as

$$x_i(t+1) = \prod_{[0,1]}(x_i(t) + u_i(t) + b_i(t))$$

Given a set $S \subseteq [0,1]^n$, we say S is reached at time t if $x(t) \in S$, and is reached in the time $[t_1, t_2]$ if there exists $t' \in [t_1, t_2]$ such that $x(t') \in S$. Based on this, we define the robust reachability of a set as follows.

Definition 5.2 Let S_1, $S_2 \subseteq [0,1]^n$ be two state sets. Under protocol (5.10) (or (5.10)), S_1 is said to be finite-time robustly reachable from S_2, if, for any $x(0) \in S_2$, S_1 is reached at time 0, or there exist constants $T>0$ and $\varepsilon \in (0,\eta)$ independent of $x(0)$ such that we can find $\delta_i(t) \in [\varepsilon, \eta)$ and $u_i(t) \in [-\eta+\delta_i(t), \eta-\delta_i(t)]$, $i \in V$, $0 \leqslant t < T$ which guarantee that S_1 is reached in the time $[1,T]$ or arbitrary $b_i(t) \in [-\delta_i(t), \delta_i(t)]$, $i \in V$, $0 \leqslant t < T$.

With this definition and similar to Lemma 3.1 in [23], we get the following lemma:

Lemma 5.1 Assume that (A1) holds. Let $S_1, \cdots, S_k \subseteq [0,1]^n$, $k \geqslant 1$ be state sets and assume they are finite-time robustly reachable from $[0,1]^n$ under protocol (5.10) (or (5.10)). Suppose the initial opinions $x(0)$ are arbitrarily given. Then for system (5.2) (or (5.3)):

(1) With probability 1 there exists an infinite sequence $t_1 < t_2 < \cdots$ such that S_j is reached at time t_{lk+j} for all $j=1,\cdots,k$ and $l \geqslant 0$.

(2) There exist constants $T>0$ and $c \in (0,1)$ such that

$$\mathbb{P}(\tau_i - \tau_{i-1} > t) \leqslant c^{\lfloor t/T \rfloor}, \quad \forall i,\ t \geqslant 1$$

where $\tau_0 = 0$ and $\tau_i = \min\{s: \text{there exist } \tau_{i-1} < t_{1'} < t_{2'} < \cdots < t_{k'} = s \text{ such that for all } j \in [1,k],\ S_j \text{ is reached at time } t_{j'}\}$ for $i \geqslant 1$.

Proof (1) This proof is similar to the proof of Lemma 3.1 (1) in [23]. To simplify the exposition we only give a proof sketch here. First, because $S_j (1 \leqslant j \leqslant k)$ is finite-time robustly reachable under protocol (5.10) (or (5.10)), so with Definition 5.2 there exist constants $T_j \geqslant 2$ and $\varepsilon_j \in (0,\eta)$ such that for any $t \geqslant 0$ and $x(t) \notin S_j$, we can find parameters $\delta_i(t') \in [\varepsilon_j, \eta)$ and control inputs $u_i(t') \in [-\eta+\delta_i(t'), \eta-\delta_i(t')]$, $1 \leqslant i \leqslant n$, $t \leqslant t' \leqslant t+T_j-2$ with which the set S_j is reached in the time $[t+1, t+T_j-1]$ for any uncertainties $b_i(t') \in [-\delta_i(t'), \delta_i(t')]$, $1 \leqslant i \leqslant n$, $t \leqslant t' \leqslant t+T_j-2$. This acts on protocol (5.2) (or (5.3)) indicating that for any $x(t-1) \in [0,1]^n$,

$$\begin{aligned} &\mathbb{P}(\{S_j \text{ is reached in} [t+1, t+T_j-1]\} \mid x(t-1)) \\ &\geqslant \mathbb{P}\Big(\bigcap_{t \leqslant t' \leqslant t+T_j-2} \bigcap_{1 \leqslant i \leqslant n} \{\xi_i(t') \in [u_i(t') - \delta_i(t'), u_i(t') + \delta_i(t')]\} \mid x(t-1)\Big) \end{aligned}$$

$$\geqslant \prod_{t'=t}^{t+T_j-2} \left[\underline{\rho} \prod_{i=1}^{n} (2\delta_i(t')) \right]$$
$$\geqslant \underline{\rho}^{T_j-1} (2\varepsilon_j)^{n(T_j-1)}$$

where the second inequality uses (A1) and the similar discussion to (11) in [23]. Set

$$E_{j,t} = \{ S_j \text{ is reached in} [t+1, t+T_j-1] \}$$

$E_t = \sum_{j=1}^{k} E_{j,t+\sum_{l=1}^{j-1} T_l}$, and $T = T_1 + T_2 + \cdots + T_k$. Similar to (13) in [23] we get

$$\mathbb{P}\left(\bigcap_{m=M}^{\infty} E_{mT}^c \right) = 0 \text{ for all } M > 0$$

which indicates that with probability 1 there exists an infinite sequence $m_1 < m_2 < \cdots$ such that $E_{m_l T}\, T$ occurs for all $l \geqslant 1$. By the definition of E_t, for each $l \geqslant 0$ we can find a time sequence $t_{lk+j} \in \left[m_l T + \sum_{p=1}^{j-1} T_p,\ m_l T + \sum_{p=1}^{j} T_p - 1 \right]$, $1 \leqslant j \leqslant k$ such that S_j is reached at time t_{lk+j}.

(2) Same as the proof of Lemma 3.1 (2) in [23]. □

Lemma 5.1 builds a connection between the noisy HK system (5.2) (or (5.3)) and the HK-control system (5.10) (or (5.10)). According to Lemma 5.1, to prove a set S is reached a. s. under a noisy HK system, we only need to design control algorithms for the corresponding HK-control system such that the set S is robustly reached in finite time. Before the proof of Theorem 5.1 we introduce some useful notions and lemmas.

For any constants $z \in [0,1]$ and $\alpha > 0$, define the set

$$S_{z,\alpha} := \left\{ (x_1, \cdots, x_n) \in [0,1]^n : \max_{i \in V} |x_i - z| < \alpha \right\} \tag{5.10}$$

Lemma 5.2 For any constants $z' \in [0,1]$ and $\alpha' > 0$, $S_{z',\alpha'}$ is finite-time robustly reachable from $[0,1]^n$ under protocols (5.10) and (5.10).

For any constant $\beta \in [0,1]$, denote the set

$$E_\beta = \left\{ (x_1, \cdots, x_n) \in [0,1]^n : \max_{i,j \in V} |x_i - x_j| \geqslant \beta \right\} \tag{5.11}$$

Lemma 5.3 If $r_{\min} < 1$ and $\eta > \max\left\{ \frac{r_{\min}}{2}, \frac{r_{\max}}{n} \right\}$, E_1 is finite-time robustly reachable from $[0,1]^n$ under protocol (5.10).

Lemma 5.4 For any constant $\varepsilon \in (0, \eta)$, let $c_\varepsilon = \min\{2\eta - 2\varepsilon, 1\}$. Then E_{c_ε} is finite-time robustly reachable from $[0,1]^n$ under protocols (5.10) and (5.10).

Lemma 5.5 Suppose $r_{\min}<1$ and $\eta>\frac{r_{\min}}{2}$. For any constant $\varepsilon>0$, let

$$c_\varepsilon=\min\left\{\frac{n\eta}{(n-1)\underline{\omega}_\eta}-\varepsilon,\ n\eta-\varepsilon,1\right\}$$

then E_{c_ε} is finite-time robustly reachable from $[0,1]^n$ under protocol (5.10).

The proofs of Lemmas 5.2 ~ 5.5 are given in Appendix A.

Proof of Theorem 5.1 By Lemma 5.2, $S_{\frac{1}{2},\frac{r_{\min}}{2}}$ is finite-time robustly reachable from $[0,1]^n$ under protocol (5.10), then by Lemma 5.1, under system (5.2) a. s. there exists a finite time t_1 such that $S_{\frac{1}{2},\frac{r_{\min}}{2}}$ is reached at time t_1, which indicates $d_V(t_1)<r_{\min}$. From this and (5.9) we get $x_1(t_1)=x_2(t_1)=\cdots=x_n(t_1)$ and so $d_V(t_1+1)\leqslant 2\eta$. Thus, if $\eta\leqslant\frac{r_{\min}}{2}$. we have $d_V(t_1+1)\leqslant r_{\min}$.

Repeating this process, we get

$$d_V(t)\leqslant 2\eta,\ \forall t>t_1 \tag{5.12}$$

for $\eta\leqslant\frac{r_{\min}}{2}$

Also, for any small constant $\varepsilon>0$, by Lemmas 5.1 and 5.4 we get a. s. $E_{2\eta-\varepsilon}$ is reached in finite time under protocol (5.2), so making ε tend to zero we have

$$\overline{d}_V\geqslant 2\eta \quad \text{a. s.} \tag{5.13}$$

For the case when $\eta\leqslant\frac{r_{\min}}{2}$, by (5.12) and (5.13) we have $\overline{d}_V=2\eta$ a. s. For the case when $\eta>\max\left\{\frac{r_{\min}}{2},\frac{r_{\max}}{n}\right\}$, by Lemmas 5.1 and 5.3 we can get $\overline{d}_V=1$ a. s.

It remains to prove (5.6). Using Lemma 5.2 again, we have $S_{\frac{1}{2},\frac{\alpha}{2}}$ is finite-time robustly reachable from $[0,1]^n$ under protocol (5.10). Also, by Lemmas 5.3 and 5.4 we have $E_{c_\alpha^1}$ is finite-time robustly reachable from $[0,1]^n$ under protocol (5.10). Combining these with Lemma 5.1 (2) yields (5.6). □

Proof of Corollary 5.1 If $\eta\leqslant\frac{r_{\min}}{2}$, by (5.4) we have $\overline{d}_V=2\eta\leqslant r_{\min}$ a. s. which means the system asymptotically reaches quasi-synchronization a. s. by Definition 5.1. if $\eta>\frac{r_{\min}}{2}$, by (5.4) we get a. s. $\overline{d}_V\geqslant 2\eta>r_{\min}$, a. s. the system asymptotically reaches quasi-synchronization a. s. if and only if $\eta \leqslant \frac{r_{\min}}{2}$.

For the case of $\eta \leqslant \frac{r_{\min}}{2}$, using (5.6) with $\alpha = 2\eta$ there are constants $\lambda_2 \in (0,1)$ and $\Lambda_2 > 0$ such that

$$\mathbb{P}(\tau > s) \leqslant (1 - \lambda_2)^{\lfloor s/\Lambda_2 \rfloor}, \ \forall s \geqslant 0$$

Also, by (5.12) but using τ instead of t_1 we have $d_V(t) \leqslant 2\eta$ for all $t > \tau$. □

Proof of Theorem 5.2 If $\eta \leqslant \frac{r_{\min}}{2}$, with the same discussion as the proof of Theorem 5.1 we can get $\overline{d}_V = 2\eta$ a. s. If $\eta > \frac{r_{\min}}{2}$, for any small constant $\varepsilon > 0$, by Lemmas 5.1, 5.4, and 5.5 we get a. s. $E_{\min}\{c_\eta^2 - \varepsilon, 1\}$ is reached in finite time under protocol (5.3), so making ε tend to zero we have $\overline{d}_V \geqslant \min\{c_\eta^2, 1\}$ a. s.

Similar to the proof of (5.6) we can get (5.8). □

Proof of Corollary 5.2 (1) Immediately from (5.7) and Definition 5.1. (2) Similar to the proof of Corollary 5.1. □

5.2.2 Proofs of Theorems 5.3 and 5.4

As a convenience, this subsection also provides some preparations to analyze the system (5.3) besides the system (5.4). Similar to Subsection 5.2.1, we construct two robust control systems, which can transform the analysis of systems (5.3) and (5.4) to the design of control algorithms. For $t \geqslant 0$, $i \in V$, and $j \in N_i(t) \setminus \{i\}$, let $\delta_i(t) \in (0, \eta)$ be an arbitrarily given real number, $u_{ji}(t) \in [-\eta + \delta_i(t), \eta - \delta_i(t)]$ denotes a control input, and $b_{ji}(t) \in [-\delta_i(t), \delta_i(t)]$ denotes the parameter uncertainty.

Similar to (5.10) and (5.10), for systems (5.3) and (5.4) we construct the control systems

$$\begin{aligned} x_i(t+1) &= \prod_{[0,1]}\left(\frac{1}{|N_i(t)|}\left[x_i(t) + \sum_{j \in N_i(t)\setminus\{i\}} (x_j(t) + u_{ji}(t) + b_{ji}(t))\right]\right) \\ &= \prod_{[0,1]}\left(x_i(t) + \frac{1}{|N_i(t)|}\sum_{j \in N_i(t)\setminus\{i\}} (u_{ji}(t) + b_{ji}(t))\right), \ \forall i \in V, \ t \geqslant 0 \end{aligned}$$

and

$$x_i(t+1) = \prod_{[0,1]}\left(\omega_i x_{\mathrm{ave}}(t) + \frac{1 - \omega_i}{|N_i(t)|}\left[x_i(t) + \sum_{j \in N_i(t)\setminus\{i\}} (x_j(t) + u_{ji}(t) + b_{ji}(t))\right]\right)$$

$$= \prod_{[0,1]} \left(x_i(t) + \frac{1-\omega_i}{|N_i(t)|} \sum_{j \in N_i(t) \setminus \{i\}} (u_{ji}(t) + b_{ji}(t)) \right), \ \forall i \in V, \ t \geqslant 0$$

respectively, where the last lines of (5.14) and (5.14) use (5.9).

Similar to Definition 5.2, we define the robust reachability for systems (5.14) and (5.14) as follows:

Definition 5.3 Let S_1, $S_2 \subseteq [0,1]^n$ be two state sets. Under protocol (5.14) (or (5.14)), S_1 is said to be finite-time robustly reachable from S_2 if: For any $x(0) \in S_2$, S_1 is reached at time 0, or there exist constants $T>0$ and $\varepsilon \in (0,\eta)$ such that we can find $\delta_i(t) \in [\varepsilon,\eta)$ and $u_{ji}(t) \in [-\eta+\delta_i(t), \eta-\delta_i(t)]$, $0 \leqslant t < T$, $i \in V$, $j \in N_i(t) \setminus \{i\}$, which guarantees that S_1 is reached in the time $[1,T]$ for arbitrary $b_{ji}(t) \in [-\delta_i(t), \delta_i(t)]$, $0 \leqslant t < T$, $i \in V$, $j \in N_i(t) \setminus \{i\}$.

Similar to Lemma 5.1 we get the following lemma:

Lemma 5.6 Assume (A2) holds. Let $S_1, \cdots, S_k \subseteq [0,1]^n$, $k \geqslant 1$ be state sets and assume they are finite-time robustly reachable from $[0,1]^n$ under protocol (5.14) (or (5.14)). Suppose the initial opinions $x(0)$ are arbitrarily given. Then for system (5.3) (or (5.4)):

(1) With probability 1 there exists an infinite sequence $t_1 < t_2 < \cdots$ such that S_j is reached at time t_{lk+j} for all $j=1,\cdots,k$ and $l \geqslant 0$.

(2) There exist constants $T>0$ and $c \in (0,1)$ such that

$$\mathbb{P}(\tau_i - \tau_{i-1} > t) \leqslant c^{\lfloor t/T \rfloor}, \ \forall i, \ t \geqslant 1$$

where $\tau_0 = 0$ and $\tau_i = \min\{s: \text{there exist } \tau_{i-1} < t_{1'} < t_{2'} < \cdots < t_{k'} = s \text{ such that for all } j \in [1,k], S_j \text{ is reached at time } t_{j'}\}$ for $i \geqslant 1$.

Similar to Lemmas 5.2, 5.4 and 5.5 we can get:

Lemma 5.7 For any constants $\bar{z} \in [0,1]$ and $\bar{\alpha} > 0$, $S_{\bar{z},\bar{\alpha}}$ is finite-time robustly reachable from $[0,1]^n$ under protocol (5.14).

Lemma 5.8 For any constant $\varepsilon \in (0,\eta)$, let $c_\varepsilon = \min\{\max_{i \neq j} a_{ij} - 2\varepsilon, 1\}$. Then E_{c_ε} is finite-time robustly reachable from $[0,1]^n$ under protocol (5.14).

Lemma 5.9 Suppose that $n \geqslant 3$. For any constant $\varepsilon > 0$, let $c_\varepsilon = \max_{i \neq j} c_{ij} - \varepsilon$, then E_{c_ε} is finite-time robustly reachable from $[0, 1]$ under protocol (5.14).

The proof of Lemma 5.7 is similar to the proof of Lemma 5.2, while the proofs of Lemmas 5.8 and 5.9 are postponed to Appendix.

Proof of Theorem 5.3 For any constant $\varepsilon \geqslant 0$, let $c_\varepsilon^* = \min\{\max_{i \neq j}\{a_{ij}, c_{ij}\}, 1\} - \varepsilon$ Combining Lemma 5.6 with Lemmas 5.8 and 5.9 we have a. s. $E_{c_\varepsilon^*}$ is reached in finite time under protocol (5.4), so making ε tend to zero we have

$$\overline{d}_V \geqslant c_0^* \quad \text{a. s.} \tag{5.14}$$

For the case that $\eta \leqslant \min\limits_{i \neq j} \frac{nr_i}{(n-1)(2-\omega_i-\omega_j)}$, this implies that $r_i \geqslant a_{ij}$ for all $i \neq j$, and then

$$c_{ij} \leqslant h_{ij} + h_{ji} = a_{ij} \leqslant r_i \leqslant 1, \ \forall i \neq j$$

By this with (5.14) we have

$$\overline{d}_V \geqslant \max_{i \neq j} a_{ij} \quad \text{a. s.} \tag{5.15}$$

Also, by Lemmas 5.6 and 5.7 we get a. s. there exists a finite time t_1 such that $|x_i(t_1) - x_j(t_1)| \leqslant a_{ij} \leqslant \min\{r_i, r_j\}$ for any $i \neq j$, which indicates $x_1(t_1) = \cdots = x_n(t_1)$. Here we recall that $x_i(t)$ is defined by (5.9). In addition

$$x_i(t_1+1) = \prod_{[0,1]}\left(x_i(t_1) + \frac{1-\omega_i}{n}\sum_{j \neq i}\zeta_{ji}(t_1)\right)$$

So

$$\begin{aligned} |x_i(t_1+1) - x_j(t_1+1)| &\leqslant \frac{1-\omega_i}{n}\sum_{k \neq i}|\zeta_{ki}(t_1)| + \frac{1-\omega_j}{n}\sum_{k \neq j}|\zeta_{kj}(t_1)| \\ &\leqslant \frac{(2-\omega_i-\omega_j)(n-1)\eta}{n} \\ &= a_{ij} \leqslant \min\{r_i, r_j\} \end{aligned}$$

Repeating this process we get $|x_i(t) - x_j(t)| \leqslant a_{ij}$ for all $t > t_1$, which implies that $\overline{d}_V \leqslant \max\limits_{i \neq j} a_{ij}$ Combining this with (5.15) we get $\overline{d}_V = \max\limits_{i \neq j} a_{ij}$ a. s. □

Proof of Theorem 5.4 For any $\alpha > 0$, by Lemmas 5.1 and 5.2 the set $S_{\frac{1}{2},\alpha}$ is reached a. s. in finite time under systems (5.2) and (5.3), so we can get $\underline{d}_V = 0$ a. s. by making α tend to zero.

Similarly, by Lemmas 5.6 and 5.7 the set $S_{\frac{1}{2},\alpha}$ is reached a. s. in finite time under systems (5.4), so we can get $\underline{d}_V = 0$ a. s. by making α tend to zero. □

5.3 Increasing Confidence Thresholds May Harm Synchronization under System (5.3)

From the study of Section 5.2, we see that small noise will lead to quasi-synchronization for the HK model with environment noise. For the HK model with communication noise, this result still holds in the homogeneous case, though it may be not true in the heterogeneous case. Interestingly, we will show that the quasi-synchronization may be broken if the confidence thresholds of constituent agents are increased.

We only give the analytic result for the homogenous case of the system (5.3). If $r_1 = r_2 = \cdots = r_n < \frac{1}{n-1}$, the values of $\overline{d}_V$ and $\underline{d}_V$ depend on the initial opinion. For example, if $x_i(0) = \frac{i-1}{n-1}$ for $1 \leqslant i \leqslant n$, then all agents are isolated and will remain unchanged, which indicates $\overline{d}_V = \underline{d}_V = 1$ for any η. However, if $|x_i(0) - x_j(0)| \leqslant r_1$ for all $i \neq j$, it can be obtained that $|x_i(t) - x_j(t)| \leqslant r_1$ for all $t \geqslant 1$ when $\eta \leqslant \frac{nr_1}{2(n-1)}$, which indicates $\underline{d}_V \leqslant \overline{d}_V \leqslant r_1$. Thus, to avoid dependence on the initial opinion, this chapter only considers the case when $r_1 = \cdots = r_n \geqslant \frac{1}{n-1}$.

First, we need introduce some lemmas as follows:

Lemma 5.10 Consider the protocol (5.14) satisfying $r_1 = \cdots = r_n \geqslant \frac{1}{n-1}$. Then for any constants $z^* \in [0,1]$ and $\alpha^* > 0$, S_{z^*,α^*} is finite-time robustly reachable from $[0,1]^n$, where S_{z^*,α^*} is the state set defined by (5.10).

Lemma 5.11 Consider the protocol (5.14) satisfying $r_1 = \cdots = r_n \in \left[\frac{1}{n-1}, 1\right)$. Let $c_\eta := \frac{2\eta(n-1)}{n}$. For any $\varepsilon > 0$, if $\eta \leqslant \frac{nr_1}{2(n-1)}$ then the set

$$E'_\varepsilon = \{(x_1, \cdots, x_n): c_\eta - \varepsilon \leqslant \max |x_i - x_j| \leqslant c_\eta\}$$

is finite-time robustly reachable from $[0,1]^n$.

Lemma 5.12 Consider the protocol (5.14) satisfying $n \geqslant 3$ and $r_1 = \cdots = r_n \in \left[\frac{1}{n-1}, 1\right)$. If $\eta > \frac{nr_1}{2(n-1)}$ then E_1 is finite-time robustly reachable from $[0,1]^n$, where E_1 is the state set defined by (5.11).

The proofs of Lemmas 5.10, 5.11, and 5.12 are postponed to Appendix.

Theorem 5.5 Consider the system (5.3) satisfying (A2), $n \geqslant 3$ and $r_1 = \cdots = r_n \in \left[\frac{1}{n-1}, 1\right)$. Then for any initial opinions $x(0) \in [0,1]^n$, a. s. $\underline{d}_V = 0$, and

$$\overline{d}_V = \begin{cases} \frac{2\eta(n-1)}{n} & \text{if } \eta \leqslant \frac{nr_1}{2(n-1)} \\ 1 & \text{otherwise} \end{cases}$$

Also, for any $\alpha > 0$ we set

$$c_\alpha = \begin{cases} \frac{2\eta(n-1)}{n} - \alpha & \text{if } \eta \leqslant \frac{nr_1}{2(n-1)} \\ 1 & \text{otherwise} \end{cases}$$

and define $\tau_0 = 0$, and τ_k to be the stopping time as

$$\tau_k = \begin{cases} \min\{s > \tau_{k-1} : d_V(s) \leqslant \alpha\}, & \text{if } k \text{ is odd} \\ \min\{s > \tau_{k-1} : d_V(s) \geqslant c_\alpha\}, & \text{if } k \text{ is even} \end{cases}$$

then for all $j \geqslant 0$ and $t \geqslant 0$

$$\mathbb{P}(\tau_{2j+2} - \tau_{2j} > t) \leqslant (1 - \lambda_5)^{\lfloor \frac{t}{\Lambda_5} \rfloor} \tag{5.16}$$

Where $\lambda_5 \in (0,1)$ and $\Lambda_5 > 0$ are constants only depending on n, η, $\underline{\rho}$, α and r_1.

Proof First by Lemmas 5.10 and 5.6 we can get a. s. $\underline{d}_V = 0$ when we let the value of α in Lemma 5.10 tend to 0.

Next, we consider the value of $\overline{d}_V$. For the case that $\eta \leqslant \frac{nr_1}{2(n-1)}$, since $\underline{d}_V = 0$ a. s. there exists a finite time t_1 such that $\max_{i,j} |x_i(t_1) - x_j(t_1)| \leqslant \frac{n-1}{n} 2\eta \leqslant r_1$. By (5.3) and (5.9) we have $\tilde{x}_1(t_1) = \cdots = \tilde{x}_n(t_1)$, and

$$x_i(t_1 + 1) \in \left[\tilde{x}_i(t_1) - \frac{n-1}{n}\eta, \tilde{x}_i(t_1) + \frac{n-1}{n}\eta\right]$$

which indicates

$$|x_j(t+1) - x_i(t+1)| \leqslant \frac{n-1}{n} 2\eta \leqslant r_1$$

Repeating the above process, we get that for any $t \geqslant t_1$,

$$\max_{i,j} |x_i(t) - x_j(t)| \leqslant \frac{n-1}{n} 2\eta \tag{5.17}$$

Combining (5.17) with Lemmas 5.11 and 5.6 we get $\overline{d}_V = \frac{2\eta(n-1)}{n}$ a. s. when we let the value of ε in Lemma 5.11 tend to 0.

If $\eta > \frac{nr_1}{2(n-1)}$, by Lemmas 5.12 and 5.6 we have a. s. $\bar{d}_V = 1$.

Finally, combining Lemma 5.6 with Lemmas 5.10, 5.11 and 5.12 yields (5.16).

□

Remark 5.4 In Theorem 5.3, if $\{\omega_i\}_{1\leqslant i\leqslant n}$ all tend to 0^+ then almost surely

$$\bar{d}_V \begin{cases} = \frac{2(n-1)\eta}{n}, & \text{if } \eta \leqslant \min_i \frac{nr_i}{2(n-1)} \\ \geqslant \min\left\{\max_{i\neq j}\left\{\frac{2(n-1)\eta}{n}, c_{ij}^*\right\}, 1\right\}, & \text{otherwise} \end{cases} \tag{5.18}$$

where c_{ij}^* is the limit value of c_{ij} as $\{\omega_i\}_{1\leqslant i\leqslant n}$ all tend to 0^+. Theorem 5.5 is consistent with (5.18) for the case when $\eta \leqslant \min_i \frac{nr_i}{2(n-1)}$, but is stronger than (5.18) for the case when $\eta > \min_i \frac{nr_i}{2(n-1)}$. Thus, Theorem 5.5 is not a special case of Theorem 5.3.

For the system (5.3), we only consider the homogeneous case since the heterogeneous case is quite difficult to analyze in detail. When considering the heterogeneous system (5.3), the finite-time robust reachability (Lemma 5.10) may not hold under some configurations. Then, in contrast to the system (5.2), in system (5.3), small noise may not lead to quasi-synchronization. More interestingly, if we increase the confidence thresholds of constituent agents, the quasi-synchronization may be broken. Also, the value of $\bar{d}_V$ as a function of η may exhibit a phase transition at some critical points. However, raising η may promote synchronization, which is different from the systems (5.2) and (5.3). To illustrate these phenomena, we give an example as follows:

Example Assume that system (5.3) contains 4 agents, and the communication noises are independently and uniformly distributed in $[-\eta, \eta]$. If $r_1 = r_2 = r_3 = r_4 = \frac{1}{3}$, for any $\eta \in \left(0, \frac{2}{9}\right)$ and any initial opinions, by Theorem 5.5 the system will reach quasi-synchronization a. s.. The evolution of opinions is shown in Fig. 5.6 with $\eta = 0.1$. However, if we increase the values of r_2 and r_3 to 1, and suppose $x(0) = \left(0, \frac{1}{2}, \frac{1}{2}, 1\right)$, then for any $\eta \in \left(0, \frac{1}{9}\right)$, it can be shown that $x_1(t) = 0$, $x_4(t) = 1$ for all $t \geqslant 1$, while $x_2(t)$ and $x_3(t)$ fluctuate between $\left(\frac{1}{2} - \frac{3}{2}\eta, \frac{1}{2} + \frac{3}{2}\eta\right)$. The evolution of opinions is shown in Fig. 5.7 with $\eta = 0.1$.

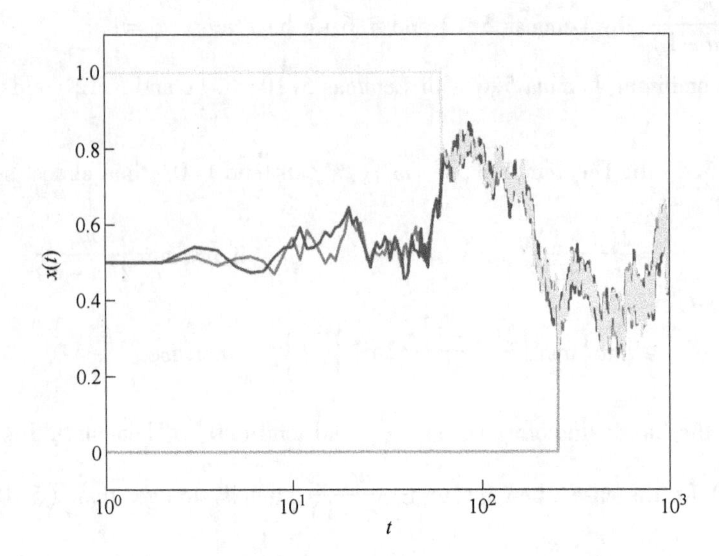

Fig. 5.6 The evolution of system (5.3) with $n=4$, $x(0)=\left(0, \frac{1}{2}, \frac{1}{2}, 1\right)$, $(r_1, \cdots, r_4)=\left(\frac{1}{3}, \frac{1}{3}, \frac{1}{3}, \frac{1}{3}\right)$, and $\eta=0.1$

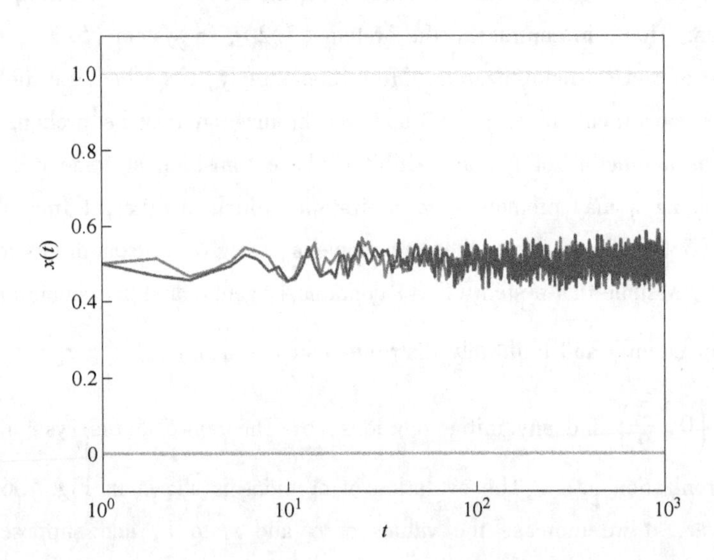

Fig. 5.7 The evolution of system (5.3) with $n=4$, $x(0)=\left(0, \frac{1}{2}, \frac{1}{2}, 1\right)$, $(r_1, \cdots, r_4)=\left(\frac{1}{3}, 1, 1, \frac{1}{3}\right)$, and $\eta=0.1$

5.4 Notes

The agreement and disagreement analysis of opinion dynamics has attracted an increasing amount of interest in recent years. On the other hand, all natural systems are inextricably affected by noise, and how noise affects the collective behavior of a complex system has also garnered considerable interest from researchers and developers in various fields. Thus, a natural problem is how the noise affects the agreement or bifurcation in opinion dynamics. This chapter analyzes heterogeneous HK models with either environment or communication noise for the first time, and provides some critical results for quasi-synchronization.

There are still some problems that have not been considered. For example, because of limited space, this chapter does not consider heterogeneous HK models with both environment and communication noises. Such systems may exhibit some interesting properties different from the systems (5.2) ~ (5.4), though the analysis may be much more complex. These problems and models leave us with a direction for future research.

References

[1] Blondel V D, Hendrickx J M, Tsitsiklis J N. On Krause's multi-agent consensus model with state-dependent connectivity[J]. IEEE Transactions on Automatic Control, 2009, 54 (11): 2586 ~ 2597.

[2] Blondel V D, Hendrickx J M, Tsitsiklis J N. Continuous-Time Average-Preserving Opinion Dynamics with Opinion-Dependent Communications[J]. SIAM Journal on Control and Optimization, 2010, 48 (8): 5214 ~ 5240.

[3] Touri B, Nedic A. Discrete-time opinion dynamics, in 2011 Conference Record of the Forty Fifth Asilomar Conference on Signals[J]. Systems and Computers (ASILOMAR), 2011.

[4] Chazelle B. The total senergy of a multiagent system[J]. SIAM Journal on Control and Optimization, 2011, 49: 1680 ~ 1706.

[5] Bhattacharyya A, Braverman M, Chazelle B, Nguyen H L. On the convergence of the hegselmann-krause system[C]. Proceedings of the 2013 ACM Conference on Innovations in Theoretical Computer Science, 2013, 61 ~ 66.

[6] Mora˘rescu I C, Girard A. Opinion dynamics with decaying confidence: Application to community detection in graphs[J]. IEEE Transactions on Automatic Control, 2011, 56 (8): 1862 ~ 1873.

[7] Motsch S, Tadmor E. Heterophilious dynamics enhances consensus[J]. SIAM Review, 2014, 56 (4): 577 ~ 621.

[8] Wedin E, Hegarty P. The Hegselmann-Krause dynamics for the continuous-agent model and a regular opinion function do not always lead to consensus[J]. IEEE Transactions on Automatic Control, 2015, 60 (9): 2416 ~ 2421.

[9] Su W, Gu Y J, Wang S, Yu Y G. Partial convergence of heterogeneous Hegselmann-Krause opinion dynamics[J]. Science China Technological Sciences, 2017, 60 (9): 1433 ~ 1438.

[10] Etesami S, Babar T. Game-theoretic analysis of the Hegselmann-Krause model for opinion dynamics in finite dimensions[J]. IEEE Transactions on Automatic Control, 2015, 60 (7): 1886 ~ 1897.

[11] Chazelle B, Wang C. Inertial Hegselmann-Krause systems[J]. IEEE Transactions on Automatic Control, 2017, 62 (8): 3905 ~ 3913.

[12] Mirtabatabaei A, Bullo F. Opinion dynamics in heterogeneous networks: convergence conjectures and theorems[J]. SIAM Journal on Control and Optimization, 2012, 50 (5): 2763 ~ 2785.

[13] Lorenz J, Heterogeneous bounds of confidence: meet, discuss and find consensus[J]. Complexity, 2010, 15 (4): 43 ~ 52.

[14] Sagués F, Sancho J M, García-Ojalvo J. Spatiotemporal order out of noise[J]. Reviews of Modern Physics, 2007, 79 (3): 829 ~ 882.

[15] Shinbrot T, Muzzio F J. Noise to order[J]. Nature, 2001, 410 (6825): 251 ~ 258.

[16] Mäs M, Flache A, Helbing D. Individualization as driving force of clustering phenomena in humans[J]. PLoS Computational Biology, 2010, 6 (10), e1000959.

[17] Pineda M, Toral R, Hernández-García E. Diffusing opinions in bounded confidence processes [J]. Eur. Phys. J. D, 2011, 62 (1): 109 ~ 117.

[18] Grauwin S, Jensen P. Opinion group formation and dynamics: Structures that last from nonlasting entities[J]. Physical Review E-Statistical, Nonlinear, and Soft Matter Physics, 2012, 85 (6), 066113.

[19] Carro A, Toral R, San Miguel M. The role of noise and initial conditions in the asymptotic solution of a bounded confidence, continuous-opinion model [J]. Journal of Statistical Physics, 2013, 151 (1 ~ 2): 131 ~ 149.

[20] Pineda M, Toral R, Hernández-García E. The noisy Hegselmann-Krause model for opinion dynamics[J]. The European Physical Journal B, 2013, 86 (12): 1 ~ 10.

[21] Su W, Chen G, Hong Y G. Noise Leads to Quasi-Consensus of Hegselmann-Krause Opinion Dynamics[J]. Automatica, 2017, 85, 448 ~ 454.

[22] Wang C, Li Q, Weinan E, Bernard C. Noisy Hegselmann-Krause Systems: Phase Transition and the 2R-Conjecture[J]. Journal of Statistical Physics, 2015, 166 (5): 1209 ~ 1225.

[23] Chen G. Small noise may diversify collective motion in Vicsek model[J]. IEEE Transactions on Automatic Control, 2017, 62 (2): 636 ~ 651.

[24] Pineda M, Buendia G M. Mass media and heterogeneous bounds of confidence in continuous opinion dynamics[J]. Physica A: Statistical Mechanics and its Applications, 2015, 420, 73 ~ 84.

[25] Chen S, Glass D H, McCartney M. Characteristics of successful opinion leaders in a bounded confidence model[J]. Physica A: Statistical Mechanics and its Applications, 2016, 449, 426 ~ 436.

[26] Chen X, Zhang X, Xie Y, Li W. Opinion Dynamics of Social-Similarity-Based Hegselmann-

Krause Model, Complexity[M]. DOI:https://doi.org/10.1155/2017/1820257, 2017.

[27] Björsell N, Händel P. Truncated Gaussian noise in ADC histogram tests[J]. Measurement: Journal of the International Measurement Confederation, 2007, 40 (1): 36 ~42.

[28] Cai G Q, Wu C. Modeling of bounded stochastic processes[J]. Probabilistic Engineering Mechanics, 2004, 19 (3): 197 ~203.

6 Random Information Flow Benefits Truth Seeking

Among the diverse issues concerned in the social dynamics, an interesting topic is to investigate how we can approach the truth in a social opinion group[1,2]. In this topic, what concerns us is what factors could be beneficial to the truth seeking. In [1], a truth seeking model was built based on the well-known Hegselmann-Krause (HK) confidence-based opinion dynamics. In this model, various factors are shown to be connected to the truth seeking of the group, including the proportion of the truth seekers, the attraction strength of truth to truth seekers, the confidence threshold of the individuals, the position of the truth value and the initial opinion values of all individuals. Nevertheless, how these factors determine the truth seeking is intricate and even counter-intuitive somehow. However, the situation is far more complex, since it can be shown that the more truth seekers and the larger attraction strength could make a reachable truth unreachable. Actually, the inherent belief inclination is nearly certain in producing the opinion cleavage[3].

Apart from the factors mentioned above, are there any other remarkable fac-tors that could affect truth seeking? For social opinion dynamics, an inevitable influence factor is the free information flow in the society, especially in this Internet era. People's opinions are affected not only by communicating with the other individuals, but also by receiving the intensive information from the free media, such as the newspapers, the TV shows, the broadcast, and especially the information flow through the social media nowadays. Usually, this free information flow can be modeled as persistent random noise added to the opinion dynamics model. By simulating and analyzing the noisy opinion models, a fascinating phenomenon was found that the random noise could play a positive role in enhancing the consensus of opinions[4~7]. A strict theoretical analysis of this fact was established based on the HK model in previous chapters. This fact reveals that free information flow could affect the social opinion evolution essentially.

With the above findings about the free information flow, we also want to know whether free information flow could benefit the truth seeking in a society. In this chapter, using the truth seeking model built in [1], we indeed find that the free information flow can effectively help the individuals acquire the truth. To be specific, it is strictly proved in this chapter that, under the influence of the random noise whose strength is below a proper value, the opinion values of all individuals almost surely approach the truth value and stay nearby, as long as there is only one active truth seeker or more in the group. The effective noise strength is shown to only depend on the intrinsic parameters of the group, i. e., the number of the truth seekers, the size of the group, the attraction strength of the truth, and the confidence threshold. Though how to approach the truth is a complicated problem which involves intricate factors, the findings in this chapter actually explicitly reveal that free information flow is a crucial element.

6.1 Preliminary and Formulation

6.1.1 Truth Seeking Model

The truth seeking model adopted in this chapter was originally established in noise-free type in [1] where the basic opinion evolution mechanism obeys the confidence-based HK dynamics. Suppose there are n individuals in a group, $[0, 1]$ is the opinion value interval where 0 and 1 denote the extreme opinions of the people towards one topic, then

$$x_i(t+1) = (1-\alpha_i)\sum_{j\in N_i(x(t))}\frac{x_j(t)}{|N_i(x(t))|} + \alpha_i A \tag{6.1}$$

for $i \in V$, where $V=\{1,\cdots,n\}$ is the group of the agents, $x_i(t)\in[0,1]$ is opinion value of agent i at time t, $\varepsilon\in[0,1]$ is the common confidence threshold, $A\in[0,1]$ is the truth value, $\alpha_i\in[0,1]$ is the attraction strength of truth for agent i, and

$$N_i(x(t)) = \{j\in V \mid |x_i(t)-x_j(t)|\leqslant\varepsilon\} \tag{6.2}$$

is the neighbor set of i at time t. Here, $|\cdot|$ represents the absolute value of a real number and the cardinality of a set accordingly.

In this model, the opinion value of each individual could be influenced by learning from its neighbors, as well as the attraction of truth. The agent i is said to be sequacious if $\alpha_i=0$, and the truth seeker otherwise. Denote $S=\{i\in V: 0<\alpha_i\leqslant 1\}$ as the set of truth seekers. When $|S|=0$, i. e., there is no truth seeker in the group, the model (6.1) degenerates to the classical Hegselmann-Krause opinion dynamics, where opinion fragmentation may happen, let alone fail in seeking the truth. When $|S|=n$, i. e., all the individuals are truth seekers. For example, intuitively, the more truth seekers exist and the stronger the strength of the truth attraction is, the more likely the group will approach the truth provided the other conditions unchanged. the truth is surely to be achieved[1]. If $1 \leqslant |S| < n$, i. e. , only part of agents are truth seekers, the truth can achieve by the truth seekers and some of their neighbors, while the other individuals cannot acquire the truth[1] (see Fig. 6. 1 in Section 6. 3 for illustration).

6.1.2 Free Information Flow

As a philosopher, Rainer Hegselmann and the coauthor elaborated the model (6. 1) and investigated the definition of the truth much from the viewpoint of philosophy and sociology in [1]. Despite their profound and professional insights about the ingredients of a system that determine truth seeking, we would modestly recognize that the stochastic information flow could also be an inevitable element in this process.

In real situations, especially in this information explosion era, people's attitudes or beliefs can be immensely influenced by the massive information that scurries in the media of the news, the broadcast, the TV shows, and the Internet. Each day, people encounter intensive news quite randomly, and this information flow affects one's opinion in a stochastic way, more or less, positive or negative. Hence, the truth seeking model (6. 1) should be added some random noises: for $t \geqslant 0$, $i \in V$,

$$x_i(t+1) = (1 - \alpha_i I_{\{i \in S\}}) \sum_{j \in N_i(x(t))} \frac{x_j(t)}{|N_i(x(t))|} + \alpha_i I_{\{i \in S\}} A + \xi_i(t+1) \tag{6.3}$$

where $I_{\{i \in S\}}$ is the indicator function, which takes value 1 or 0, according to $i \in S$ or not. Furthermore, in real situations, how much and in what manner the free information flow affects the individuals' opinions are quite intricate and untraceable, hence for the random noises $\{\xi_i(t)\}_{i \in V, t > 0}$ which model the free information flow, we bluntly but appropriately assume that they are the independent random variables uniformly distributed on $[-\delta, \delta]$ with $\delta > 0$.

On account of the extreme opinions in reality, in our model settings in the following, we suppose that the noisy opinion values are still limited in a closed interval, say $[0,1]$ without loss of generality. Also, in our analysis, we only consider the homogenous attraction strength for convenience, i. e. , $\alpha_i = \alpha \in (0,1]$, $i \in S$, then for $t \geqslant 0$,

$$x_i(t+1) = \begin{cases} 1, & x_i^*(t) > 1 \\ x_i^*(t), & x_i^*(t) \in [0, 1], \ \forall i \in V \\ 0, & x_i^*(t) < 0 \end{cases} \tag{6.4}$$

where

$$x_i^*(t) = (1 - \alpha I_{\{i \in S\}}) \sum_{j \in N_i(x(t))} \frac{x_j(t)}{|N_i(x(t))|} + \alpha I_{\{i \in S\}} A + \xi_i(t+1) \quad (6.5)$$

and $1 \leqslant |S| \leqslant n$,

$$\{\xi_i(t), i \in V, t > 0\} \text{ are independent and uniformly distributed on} [-\delta, \delta] \quad (6.6)$$

6.1.3 Approach Truth

For the truth seeking model (6.1), the system can be said to achieve the truth A if $\lim_{t \to \infty} x_i(t) = A$, $i \in V$. However, with the presence of random noise, which could fluctuate the opinions with the amplitude of the noise strength, we should give a definition for the system (6.4) ~ (6.6) to achieve the truth approximately:

Definition 6.1 Denote $d_V^A(t) = \max_{i \in V} |x_i(t) - A|$, then for $\phi \geqslant 0$, we say the system (6.4) ~ (6.6) will approach the truth with ϕ - precision, if $\limsup_{t \to \infty} d_V^A(t) \leqslant \phi$ holds.

6.2 Main Results

As Fig. 6.1 shows, it is easy to check that, unless all the individuals are truth seekers in the group, there always exist initial conditions that the system cannot achieve the truth. However, in the following we will strictly prove that, with the free information flow, the system could approach the truth almost surely given any initial conditions.

To begin with and to simplify the narration, some denotations are introduced. Denote $m = |S|$ be the number of the truth seekers and

$$\begin{aligned} &\delta_1 = \frac{n(1-\alpha)\delta}{m\alpha} + \delta, \ \delta_2 = \frac{n\delta}{m\alpha} + \delta, \ \bar{\delta} = \max\{\delta_1, \delta_2\} \\ &\underline{\delta} = \min\left\{\frac{m\alpha}{2n + (2m-n)\alpha}\varepsilon, \ \frac{m}{n+2m}\varepsilon\right\} \end{aligned} \quad (6.7)$$

where n, α, δ, ε are the intrinsic parameters collected in the previous introduction of the truth seeking model (6.2) and (6.4) ~ (6.6), and let a. s. be the abbreviation of almost surely by which we say an event B occurs a. s. if $\mathbb{P}\{B\} = 1$ in probability theory, then we have

Theorem 6.1 For any initial state $x(0) \in [0,1]^n$, $\varepsilon \in (0,1]$ and $\delta \in (0, \underline{\delta}]$, the system (6.4) ~ (6.6) will a. s. approach the truth with $\bar{\delta}$ – precision.

Remark 6.1 Here, $\bar{\delta}$ measures the precision of the truth seeking, and $\underline{\delta}$ defines the effective noise strength bound. The allowed noise strength bound depends completely on the intrinsic parameters of the system, and they together deter-mine the consequential precision of the truth seeking. It can be seen from (6.7) that the precision $\bar{\delta}$ is usually larger than the noise strength δ, and the excess amount is also determined by the actual intrinsic parameters of the system.

Remark 6.2 The truth seeking of the model (6.1) is dependent upon the "intrinsic parameters", i. e., ε, α, n, m, of the system, however, this relationship is quite intricate. However, with free information flow, the relationship between the parameters and the truth seeking result, $\bar{\delta}$ and $\underline{\delta}$, is much intuitive. Roughly, the larger m and α the system possesses, the larger the effective noise strength $\underline{\delta}$ is allowed, and the better the precision $\bar{\delta}$ is obtained for given δ.

Remark 6.3 Though Theorem 6.1 asserts that the group will almost surely approach the truth under the stirring of free information flow, in practical observation, this process could be inconceivably long lasting (refer to Fig. 6.2). Nevertheless, this result indubitably indicates that free information flow could accelerate the acquisition of the truth, especially for an isolated society with collective unconscious.

The following Corollary provides a specific case of Theorem 6.1 for a better demonstration:

Corollary 6.1 Suppose $m = \left\lceil \frac{n}{2} \right\rceil$, $\alpha = 0.5$, then a. s. $\limsup\limits_{t\to\infty} d_V^A(t) \leqslant 5\delta$ for all $\delta \in \left(0, \frac{\varepsilon}{8}\right]$.

Corollary 6.1 tells that when half components of a society are truth seekers, the group will finally approach the truth with the precision of at least 5δ when the fluctuation magnitude δ caused by free information flow is restrained in a minor limitation.

Now we turn to the proof of Theorem 6.1, and some preliminary lemmas are needed.

Lemma 6.1 Denote $d_S^A(t) = \max\limits_{i\in S} |x_i(t) - A|$, $d_{\bar{S}}^A(t) = \max\limits_{i\in \bar{S}} |x_i(t) - A| (\bar{S} = V - S)$ for $t \geqslant 0$. For $0 < \alpha \leqslant 1$, if there is a finite time T a. s. such that $d_S^A(T) \leqslant \delta_1$, $d_{\bar{S}}^A(T) \leqslant \delta_2$, then a. s. $\limsup\limits_{t\to\infty} d_S^A(t) \leqslant \delta_1$, $\limsup\limits_{t\to\infty} d_{\bar{S}}^A(t) \leqslant \delta_2$ for $0 < \delta \leqslant \underline{\delta}$.

Proof At time T, we have a. s.

$$\max_{i,j\in V} |x_i(T) - x_j(T)| \leqslant \max_{i\in V} |x_i(T) - A| + \max_{j\in V} |x_j(T) - A| \leqslant \delta_1 + \delta_2 \leqslant \varepsilon \tag{6.8}$$

This means all agents are neighbors to each other at step T, implying $|N_i(x(T))| =$

n, $i \in V$. Hence by (6.4) and (6.5),

$$\begin{aligned} x_i(T+1) &= \alpha A + (1-\alpha)\frac{\sum_{k\in V} x_k(T)}{n} + \xi_i(T+1),\ i \in S \\ x_j(T+1) &= \frac{\sum_{k\in V} x_k(T)}{n} + \xi_j(T+1),\ j \in \bar{S} \end{aligned} \tag{6.9}$$

Then a. s.

$$\begin{aligned} |x_i(T+1) - A| &= \left| \alpha A - A + (1-\alpha)\frac{\sum_{k\in V} x_k(T)}{n} + \xi_i(T+1) \right| \\ &\leqslant \frac{1-\alpha}{n}\sum_{k\in V} |A - x_k(T)| + \delta \\ &\leqslant (1-\alpha)\frac{m\delta_1 + (n-m)\delta_2}{n} + \delta \leqslant \delta_1,\ i \in S \end{aligned} \tag{6.10}$$

and a. s.

$$\begin{aligned} |x_j(T+1) - A| &= \left| \frac{\sum_{k\in V} x_k(T)}{n} - A + \xi_i(T+1) \right| \\ &\leqslant \frac{1}{n}\sum_{k\in V} |A - x_k(T)| + \delta \\ &\leqslant \frac{m\delta_1 + (n-m)\delta_2}{n} + \delta \leqslant \delta_2,\ j \in \bar{S} \end{aligned} \tag{6.11}$$

This means

$$d_S^A(T+1) \leqslant \delta_1,\ d_{\bar{S}}^A(T+1) \leqslant \delta_2$$

Repeating the above procedure, we will obtain the conclusion. □

Now we give the proof of Theorem 6.1:

Proof of Theorem 6.1 It is easy to know that for all $i \in V$, $t \geqslant 1$,

$$\mathbb{P}\left\{\xi_i(t) \in \left[\frac{\delta}{2}, \delta\right]\right\} = \mathbb{P}\left\{\xi_i(t) \in \left[-\delta, -\frac{\delta}{2}\right]\right\} = \frac{1}{4} \tag{6.12}$$

Denote $x_i(t) = \frac{1}{|N_i(x(t))|}\sum_{j\in N_i(x(t))} x_j(t)$, $t \geqslant 0$, and this denotation remains valid for the rest of the context. If $d_V^A(0) \leqslant \delta$, the conclusion holds by Lemma 6.1 since $\delta_1 \geqslant \delta$,

$\delta_2 \geqslant \delta$. Otherwise, consider the following protocol: for all $i \in V$, $t > 0$,

$$\begin{cases} \xi_i(t+1) \in \left[\frac{\delta}{2}, \delta\right], & \text{if } x_i(t) \leqslant A \\ \xi_i(t+1) \in \left[-\delta, -\frac{\delta}{2}\right], & \text{if } x_i(t) > A \end{cases} \tag{6.13}$$

By Lemma 3.1, (6.4) and (6.5), we know under the protocol (6.13) that

$$d_V^{A(1)}(1) \leqslant \max_{i \in V}\left\{ \left| x_i(0) - A - \frac{\delta}{2} \right| \right\} \leqslant \max_{i \in V}\left\{ |x_i(0) - A| - \frac{\delta}{2} \right\} \leqslant d_V^A(0) - \frac{\delta}{2}$$

By (6.12) and independence of the random noises $\{\xi_i(t), i \in V, t \geqslant 1\}$, we know that the probability of the occurrence of the protocol (6.13) at $t = 1$ is $\frac{1}{4}$, that is

$$\mathbb{P}\left\{ d_V^A(1) \leqslant d_V^A(0) - \frac{\delta}{2} \right\} \geqslant \frac{1}{4^n} \tag{6.14}$$

Let $L = \frac{1-\delta}{\delta/2}$, if $d_V^A(1) \leqslant \delta$, the conclusion holds. Otherwise, continue the above procedure L times and we can get by (6.14) and independence that

$$\mathbb{P}\{ d_V^A(L) \leqslant \delta \} \geqslant \mathbb{P}\{ \text{protocol (6.13) occurs } L+1 \text{ times} \} \geqslant \frac{1}{4^{n(L+1)}} > 0 \tag{6.15}$$

Thus

$$\mathbb{P}\{ d_V^A(L) > \delta \} \leqslant 1 - \frac{1}{4^{n(L+1)}} \tag{6.16}$$

Denote events

$$\begin{aligned} E_0 &= \Omega \\ E_m &= \{ \omega: d_V^A(t) > \delta, (m-1)L < t < mL \}, \ m \geqslant 1 \end{aligned} \tag{6.17}$$

Since $x(0)$ is arbitrarily given, by (6.16), we can get for $m \geqslant 1$ that

$$\mathbb{P}\left\{ E_m \middle| \bigcap_{j<m} E_j \right\} \leqslant \mathbb{P}\left\{ d_V^A(mL) > \delta \middle| \bigcap_{j<m} E_j \right\} \leqslant 1 - \frac{1}{4^{n(L+1)}} < 1 \tag{6.18}$$

Note by Lemma 6.1 that $\left\{ \limsup_{t \to \infty} d_V^A(t) > \bar{\delta} \right\} \subset \left\{ \bigcap_{m \geqslant 1} \{ d_V^A(t) > \delta, \ (m-1)L < t < mL \} \right\}$, then by (6.18)

$$
\begin{aligned}
\mathbb{P}\left\{\limsup_{t\to\infty} d_V^A(t) \leqslant \bar{\delta}\right\} &= 1 - \mathbb{P}\left\{\limsup_{t\to\infty} d_V^A(t) > \bar{\delta}\right\} \\
&\geqslant 1 - \mathbb{P}\left\{\bigcap_{m\geqslant 1}\{d_V^A(t) > \delta,\ (m-1)L < t < mL\}\right\} \\
&= 1 - \mathbb{P}\left\{\bigcap_{m\geqslant 1} E_m\right\} \\
&= 1 - \mathbb{P}\left\{\lim_{m\to\infty}\bigcap_{j=1}^{m} E_j\right\} = 1 - \lim_{m\to\infty}\mathbb{P}\left\{\bigcap_{j=1}^{m} E_j\right\} \\
&= 1 - \lim_{m\to\infty}\prod_{j=1}^{m}\mathbb{P}\left\{E_j \,\middle|\, \bigcap_{k<j} E_k\right\} \\
&\geqslant 1 - \lim_{m\to\infty}\left(1 - \frac{1}{4^{n(L+1)}}\right)^m = 1
\end{aligned}
$$

Here, the exchangeability of the probability and the limit holds since $\left\{\bigcap_{j=1}^{m} E_j,\ m\geqslant 1\right\}$ is a decreasing sequence and $\mathbb{P}$ is a probability measure (refer to Corollary 1.5.2 of [10]). This completes the proof. □

6.3 Simulations

In this part, we present some simulations under the guide of Corollary 6.1 to demonstrate the truth seeking with the influence of free information flow. Let $n = 20$, $x_i(0)$, $i \in V$ be randomly generated from the interval $[0, 1]$, $\varepsilon = 0.2$, $A = 0.8$, $\alpha = 0.5$ and $m = 10$. Fig. 6.1 shows the case of truth seeking when there is no free information flow ($\delta = 0$). It can be seen that only part of the individuals achieves the truth. If injecting the free information flow into the group, Corollary 6.1 tells that when the noise strength satisfies $0 < \delta \leqslant \frac{\varepsilon}{8}$, the system will approach the truth. Take the same intrinsic parameters as in Fig. 6.1 and $\delta = 0.02 = 0.1\varepsilon$, then Fig. 6.2 explicitly shows the result.

At the end, we would like to give some intuitive explanations about how free information flow enables the group to approach the truth. There are two keys. Firstly, the group has to possess the intrinsic ability to acquire and maintain the truth once it has an appropriate condition. That is, an attraction region exists for the system to maintain the desired property. Secondly, the persistently accidental fluctuations could drive the states to access the attraction region. As for the model (6.4) ~ (6.6), the existence of truth seekers (even if only one truth seeker would exist) is responsible for the former, and random noise is responsible the latter. Fig. 6.3, re-plotted from Fig. 6.2 by the "Semilogx" function of MATLAB, clearly displays the procedure.

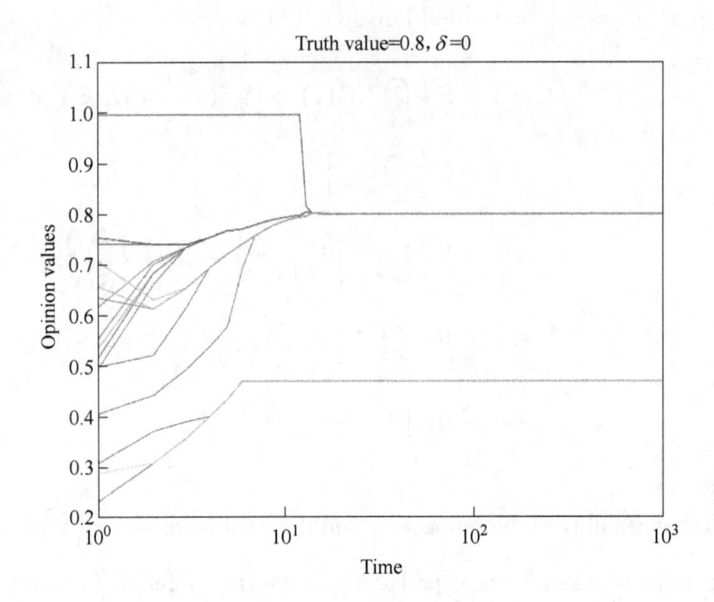

Fig. 6. 1 The truth seeking of the model (6. 1)

(Take $n=20$, $m=10$, $\alpha=0.5$, confidence threshold $\varepsilon=0.2$)

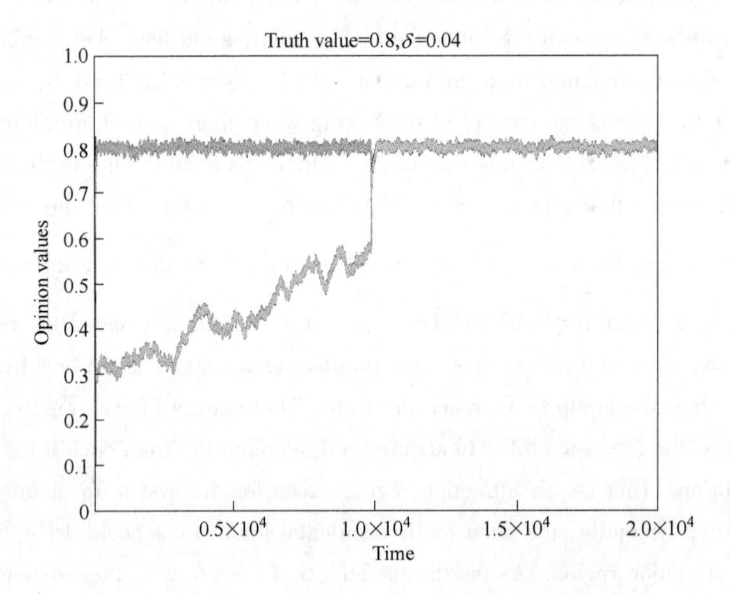

Fig. 6. 2 The truth seeking of the model (6. 4) ~ (6. 6)

(Take the same parameters and add random noise with strength $\delta=0.02$ here)

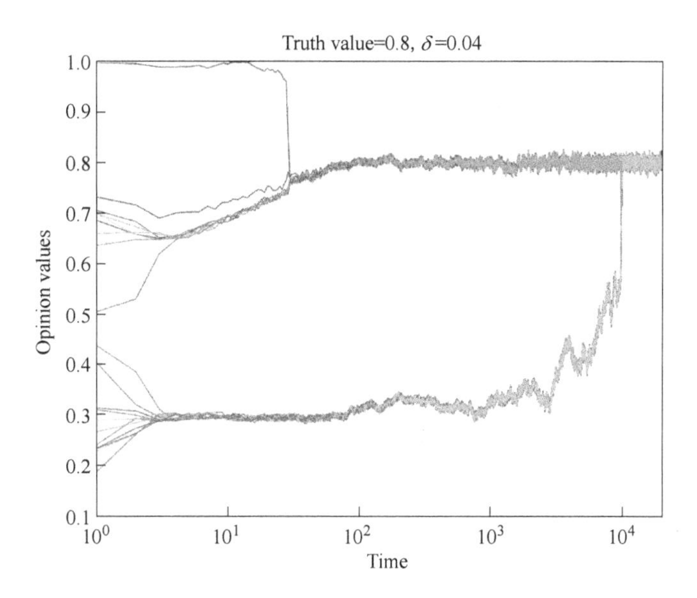

Fig. 6.3 Fig. 6.2 is replotted to better display the initial picture of evolution

6.4 Notes

This chapter used a well-established truth seeking model to theoretically reveal that free information flow could effectively enable the approach of the truth in a group. In this chapter, as well as in some previous studies, random noise has been verified to be favorable in finding and establishing the order of the system to some extent. As free information flow is ubiquitous in this information society, its essential role in determining the rich social dynamics remains to be further interpreted.

The main theorem shows that the group will finally acquire the truth when the random noise strength is small in some sense. We may declare that the feasible noise strength obtained here is somehow conservative.

Intuitively foreseeable that the excessively strong noise could destroy the truth seeking. In[8], it has been strictly proved that the consensus behavior could break down when the noise strength persistently exceeds a critical value. Fortunately, in most open societies, the fluctuation produced by free information flow is appropriately tender that the truth can be expected to be approached.

In this work, the truth seeking model is based on the HK opinion dynamics, where the opinions are updated simultaneously at each step, which may not capture the complete reality. Some complementary opinion dynamics based on bounded confidence, such as Deffuant-Weisbuch model[11~13], and some other interaction types[14], should be employed in future to investigate how the information flow determines the truth seeking of opinion dynamics.

The last issue to be concerned is to enlarge the application of the truth seeking model with the free information flow. The exploration of the definition of truth can be found in [1], where truth can be regarded as the rational tendency of the individual from another aspect. In real world, people's opinions often present binary opposition, such as leftwing and rightwing in politics. In an isolated society, an individual may hold an accordant standpoint of the authority even though he may possess an instinctive opposite tendency. In chapter 8, we may investigate how free information flow affects the opinion dynamics in this situation.

References

[1] Hegselmann R, Krause U. Truth and cognitive division of labour first steps towards a computer aided social epistemology[J]. Journal of Artificial Societies and Social Simulation, 2006, 9 (3): 1 ~28.

[2] Malarz K. Truth seekers in opinion dynamics models[J]. International Journal of Modern Physics C, 2006, 17 (10): 1521 ~1524.

[3] Friedkin N. The problem of social control and coordination of complex systems in sociology: a look at the community cleavage problem[J]. Control Systems Magazine, 2015, 35 (3): 40 ~51.

[4] Mäs M, Flache A, Helbing D. Individualization as driving force of clustering phenomena in humans[J]. PLoS Computational Biology, 2010, 6 (10): e1000959.

[5] Grauwin S, Jensen P. Opinion group formation and dynamics: structures that last from nonlasting entities[J]. Physical Review E-Statistical, Nonlinear, and Soft Matter Physics, 2012, 85 (6): 066113.

[6] Carro A, Toral R, Miguel M. The role of noise and initial conditions in the asymptotic solution of a bounded confidence, continuous-opinion model[J]. Journal of Statistical Physics, 151 (1 ~2): 131 ~149, 2013.

[7] Pineda M, Toral R, Hernndez-Garca E. The noisy Hegselmann-Krause model for opinion dynamics[J]. The European Physical Journal B, 2013, 86 (12): 1 ~10.

[8] Su W, Chen G, Hong Y. Noise leads to quasi-consensus of Hegselmann-Krause opinion dynamics [J]. Automatica, 2017, 85: 448 ~454.

[9] Blondel V D, Hendrickx J M, Tsitsiklis J N. On Krause's multi-agent consensus model with state ~ dependent connectivity[J]. IEEE Transactions on Automatic Control, 2009, 54 (11): 2586 ~ 2597.

[10] Chow Y, Teicher H. Probability Theory: Independence, Interchangeability, Martingales[M]. Springer Science & Business Media, 1997.

[11] Deffuant G, Neau D, Amblard F, Weisbuch G. Mixing beliefs among interacting agents[J]. Advances in Complex Systems, 2000, 3 (01n04): 87 ~ 98.

[12] Zhang J, Hong Y. Opinion evolution analysis for short-range and long-range Deffuant-Weisbuch models[J]. Physica A: Statistical Mechanics and its Applications, 2013, 392 (21): 5289 ~ 5297.

[13] Zhang J, Hong Y. Fluctuation analysis of a long-range opinion dynamics[C]. Proceedings of the IEEE Conference on Decision and Control, 2014, 2106 ~ 2111.

[14] Hu J, Zheng W X. Emergent collective behaviors on competition networks[J]. Physics Letters A, 2014, 378 (26 - 27): 1787 ~ 1796.

7 Elimination of Disagreement via Covert Noise

In social opinion area, one of the central issues in study of social opinions is the consensus or agreement of opinion dynamics. However, both practical observations and analysis of opinion dynamics models show that it is very often that opinions evolve to be disagreement and form a divisive status. It is explicitly known that strong cleavage may cause social conflict and subsequent catastrophe in society[1,2]. Hence, it is usually necessary to reduce opinion difference through some intervention measures. Furthermore, it sometimes needs active opinion intervention to achieve some social goals related to, e. g. , opening new market for corporations, commercial negotiations, political elections, and social campaigns, etc. [3~5], where opinions are expected to be guided to some targeted objectives. In consequence, design of simple and efficient intervention strategies to eliminate disagreement and induce opinions towards a targeted value is of great importance in opinion dynamics.

Usually, it is difficult and costly to design applicable interventions that could effectively eliminate opinion difference of a group. Fortunately, in previous chapters it is proved that when all agents with HK dynamics are affected by random noises, their opinions will almost surely reach quasi-consensus (a concept of consensus defined for the noisy case) in finite time and a critical value of noise strength is given. Since noise is quite easy to generate and manipulate (for example, injecting or spreading purposive or free information), these encouraging findings prompt us to design the disagreement elimination strategy by using random noise. However, the synchronized opinions in Chapters 3 and 4 are proved to cross the whole opinion space infinite times, while on the other hand we practically need opinions evolve to some targeted value. What's more, all agents are controlled by noise in previous chapters, which is almost impossible in a large social group and further detailed conclusion of consensus time is far from clear. In consequence, more practical and applicable noise intervention strategies require to be further explored beyond what is provided in previous chapters.

The main motivation of this chapter is to overcome the shortage of tradition control in social system, and design a feasible and low-cost noise-based control strategy to eliminate the disagreement of large social group. We intentionally introduce random noise uniformly distributed on an interval to only one targeted agent in a divisive system to affect its opinion values. Furthermore, inspired by soft control technique proposed in [6,7], where a special "shill" is introduced to control the collective behavior intentionally, here we introduce a leader agent who sticks to its opinion value in the system to intervene the objective opinions. The divisive system is established based on the HK dynamics since HK model presents a typical disagreement of opinions. For the sake of practical convenience, we only consider to inject weak noise which is covert in manipulation. The noise can be selected information or standpoints spread intentionally. Under this scheme, it will be first proved that the separated opinions almost surely reach ϕ-consensus (a general concept of consensus) in finite time when the driving noise is adequately weak, then proved that in the presence of leader agent, the synchronized opinions can run to the leader's opinion value in finite time. Especially, it is at last calculated the finite time, which is a new stopping time, when opinions reach ϕ-consensus and the objective value. It will be shown that the finite stopping time is not integrable when the noise is neutral, and integrable when the noise is oriented (integrable means the expectation of stopping time is finite). This provides us with further guides for designing more effective noise intervention strategies to eliminate disagreement of a group. Anyway, though far from perfect, this chapter takes a first step to quantitatively control opinion by using noise.

7.1 Preliminaries and Formulation

In this section, it is first introduced the divisive opinion model based on HK dy namics, then presented the noise intervened model, and proposed a "consensus" definition in noisy case at last.

7.1.1 HK-Based Divisive System

It is widely known that the HK opinion dynamics provides a typical demonstration of opinion disagreement. Suppose there are n agents with $V = \{1, \cdots, n\}$, whose opinion values at time t are denoted by $x(t)$ taking values in $(-\infty, \infty)^n$, then the basic HK model admits the following dynamics:

$$x_i(t+1) = \frac{1}{|N_i(x(t))|} \sum_{j \in N_i(x(t))} x_j(t), \quad i \in V \tag{7.1}$$

where

$$N_i(x(t)) = \{j \in V \mid |x_j(t) - x_i(t)| \leqslant \varepsilon\} \tag{7.2}$$

is the neighbor set of i and $\varepsilon > 0$ is the confidence threshold of agents. $|\cdot|$ can be the cardinal number of a set or the absolute value of a real number accordingly.

The noise-free HK model (7.1) has been proved to reach static state in finite time, with agreement or disagreement status:

Proposition 7.1[8] For every $1 \leqslant i \leqslant n$ of (7.1), $x_i(t)$ will converge to some x_i^* in finite time, and either $x_i^* = x_j^*$ or $|x_i^* = x_j^*| > \varepsilon$ holds for any i, j.

If $x_i^* = x_j^*$ for all i, j, it means all opinions reach consensus and the group realizes agreement. Otherwise, the opinions get separated and the group is divisive. In other words, a divisive opinion system with N_c distinct standpoints x_i^* $(1 \leqslant i \leqslant N_c)$ can be assumed to satisfy the following conditions:

$$\begin{cases} V = \bigcup_{g=1}^{N_c} V_g, & 2 \leqslant N_c \leqslant n \\ x_i(0) = x_g^* \in (-\infty, \infty), & \text{for } i \in V_g,\ 1 \leqslant g \leqslant N_c \\ x_p^* - x_q^* > \varepsilon, & \text{for } 1 \leqslant q < p \leqslant N_c \end{cases} \tag{7.3}$$

where $\{V_g, 1 \leqslant g \leqslant N_c\}$ are the separated subgroups of the divisive system at the initial time with $V_{g1} \cap V_{g2} = \varnothing$, $g_1 \neq g_2$ and the agents are numbered through ordering the initial states $x_1(0) \leqslant x_2(0) \leqslant \cdots \leqslant x_n(0)$ without loss of generality.

7.1.2 Noise Intervened System

An important goal in research of opinion dynamics is to enhance the consensus of opinions with some control or intervention strategies. However, it seems far from easy to implement a simple and practicable method in social intervention. The traditional controller based on the states of the system is hardly feasible for the opinion dynamics since it is almost impossible to acquire the opinion values of all agents and furthermore, the interaction topology of the highly nonlinear opinion dynamics is badly dependent on the states. Though it has been proved that random noise could synchronize the opinions, the noises therein are admitted to all agents at the beginning, and this elegant theoretical result still has limitation in applying to design a practicable method to eliminate the opinion difference, since it is hard to guarantee all opinions to be infected by noises. Moreover, consensus time and hence what noise is more effective is unknown at all.

To eliminate the disagreement of system (7.1) ~ (7.3) and also reduce the control cost, this paper tries to inject random noises to only one agent, say agent 1. To be specific, the noise intervention model of (7.1) ~ (7.3) is as follows: for $i \in V$,

$$x_i(t+1) = \frac{1}{|N_i(x(t))|}\sum_{j\in N_i(x(t))} x_j(t) + I_{\{i=1\}}\xi(t+1) \tag{7.4}$$

where the noises $\{\xi(t), t\geqslant 1\}$ are independent and uniformly distributed on

$$[\delta_1, \delta_2],\ 0\leqslant\delta_1\leqslant\delta_2<\infty \tag{7.5}$$

and $I_{\{\cdot\}}$ is the indicator function with $I_{\{i=1\}}=1$ when $i=1$ and $I_{\{i=1\}}=0$ otherwise. If there is no noise intervention ($\delta_2=0$), model (7.4) degenerates to the original one (7.1), and the system stays in divisive status with N_c separated opinions.

In practice, the random noises can be generated by exposing agent 1 to the selected information. The information can be neutral ($\delta_1=\delta_2$) or oriented ($\delta_1\neq\delta_2$). We take assumption (7.5) on the ground that the noise intervention is added to agent 1 (or any agent in cluster V_1) whose opinion value is smallest according to (7.3), then intuitively the workable noise should play a nonnegative role in general to affect the opinion. If intervention is added to agent n (or any agent in cluster V_{Nc}), the noise strength can be assumed as $0\leqslant\delta_2\leqslant\delta_1$.

7.1.3 ϕ-Consensus

To study the behavior of model (7.2) ~ (7.5), a definition of consensus in noisy case is needed:

Definition 7.1 Define

$$d_V(t) = \max_{i,j\in V}|x_i(t)-x_j(t)| \text{ and } d_V = \limsup_{t\to\infty} d_V(t)$$

For $\phi\in[0,\infty)$,

(1) If $d_V\leqslant\phi$, we say the system (7.2) ~ (7.5) will reach ϕ-consensus.

(2) If $\mathbb{P}\{d_V\leqslant\phi\}=1$, we say almost surely (a. s.) the system (7.2) ~ (7.5) will reach ϕ-consensus.

(3) Let $T=\inf\{t: d_V(t')\leqslant\phi \text{ for all } t'\geqslant t\}$. If $\mathbb{P}\{T<\infty\}=1$, we say a. s. the system (7.2) ~ (7.5) reaches ϕ-consensus in finite time.

For HK dynamics, when noise is bounded, ϕ-consensus is naturally implied by quasi-consensus.

7.2 ϕ-Consensus of Noise-Intervened System

In this part, we present the main result that the noise-intervened system (7.2) ~ (7.5) will reach ϕ-consensus in finite time:

Theorem 7.1 For any $\varepsilon > 0$ and $0 < \delta_2 \leqslant \frac{\varepsilon}{2n}$, the noise-intervened system (7.2) ~ (7.5) will a. s. reach δ_2-consensus in finite time.

There are two remarks about Theorem 7.1 compared to the main result in previous chapters. First, in previous chapters except Chapter 4, the initial opinion values $x(0)$ are arbitrarily given, but limited in a bounded interval $[0,1]^n$, while the initial opinion values here are assumed in advance, but extended to an infinite interval. In addition, there is only one agent that is infected by noise here instead of all agents. These variations cause some new mathematical challenges that the analysis skill in previous chapters fails in proving this theorem, and here it need develop a completely new method.

Second, the feasible noise strength $0 < \delta_2 \leqslant \frac{\varepsilon}{2n}$ here is quite conservative compared to that in Chapters 3 and 4, where $0 < \delta_2 \leqslant \frac{\varepsilon}{2}$ was obtained. But this result is still valuable for practical application, since it is covert and hence more feasible to use tiny noise, such as some minor disturbed information or free information flow, to eliminate the opinion difference. Especially, for small groups, this result possesses its own value. Actually, we conjecture that the effective noise strength here can be extended to $0 < \delta_2 \leqslant \varepsilon$ which is also critical as in Chapters 3 and 4. However, the existing analysis methods seem incapable of completing this.

To prove Theorem 7.1, the following lemmas are needed:

Lemma 7.1 For the system (7.2) ~ (7.5), if $0 < \delta_2 \leqslant \varepsilon$ and there exists a finite time $T \geqslant 0$ such that $d_V(T) \leqslant \varepsilon$, then $d_V(t) \leqslant \delta_2$, $t \geqslant T$ on $\{T < \infty\}$.

Proof At time T, all agents are neighbors to each other, then by (7.4), it has $x_i(T+1) = \frac{1}{n}\sum_{j=1}^{n} x_j(T) + I_{\{i=1\}}\xi(T+1)$, implying $d_V(T+1) = |\xi(T+1)| \leqslant \delta_2$, a. s.

Repeating the procedure yields the conclusion. □

Lemma 7.1 gives the invariant zone where consensus cannot be destroyed by noise.

Lemma 7.2 Still with conditions given in Lemma 7.1, then on $\{T < \infty\}$ it a. s. occurs

$$\limsup_{t \to \infty} x_i(t) = +\infty, \ i \in V$$

Proof Suppose $T = 0$ a. s. without loss of generality, then

$$
\begin{aligned}
x_i(t+1) &= \frac{1}{n}\sum_{j=1} x_j(t) + I_{\{i=1\}}\xi(t+1) \\
&= \frac{1}{n}\sum_{j=1}^{n}\left(\frac{1}{n}\sum_{k=1}^{n} x_k(t-1) + \xi(t)\right) + I_{\{i=1\}}\xi(t+1) \\
&= \frac{1}{n}\sum_{j=1}^{n} x_j(t-1) + \frac{1}{n}\xi(t) + I_{\{i=1\}}\xi(t+1) \\
&= \frac{1}{n}\sum_{j=1}^{n} x_j(0) + \sum_{k=1}^{t}\frac{1}{n}\xi(k) + I_{\{i=1\}}\xi(t+1) \qquad (7.6)
\end{aligned}
$$

Let $S_t = \frac{1}{n}\sum_{k=1}^{t-1}\xi(k)$, $t \geqslant 2$, then by (7.6)

$$
x_i(t+1) = \frac{1}{n}\sum_{j=1}^{n} x_j(0) + S_t + I_{\{i=1\}}\xi_i(t+1) \qquad (7.7)
$$

Since $\{\xi(t), t \geqslant 1\}$ are i. i. d. random variables with $\mathbf{E}\xi(1) = \frac{\delta_2 - \delta_1}{2}$, by Law of Large Number, $\frac{S_t}{t} \to \frac{\delta_2 - \delta_1}{2n}$ a. s. If $\delta_1 < \delta_2$, we have $\limsup_{t\to\infty} S_t \to \infty$ a. s. Otherwise, if $\delta_1 = \delta_2$, it is from Theorem 5. 4. 3 of[9] that $\limsup_{t\to\infty} S_t \to \infty$ a. s. Then by (7. 7), $\limsup_{t\to\infty} x_i(t) \to \infty$ a. s. □

Lemma 7. 1 states that once all opinions locate within the region with size of confidence threshold, the noise with strength no larger than ε cannot separate them anymore. Actually, ε is the upper bound of the noise strength that the noisy opinion dynamics can maintain δ_2-consensus. Using the similar skill presented in Chapters 3 and 4, we are able to obtain that the opinions will almost surely diverge at some moment when $\delta_2 > \varepsilon$. Lemma 7.2 says that once the system achieves δ_2-consensus, the opinions driven by random noises can enter arbitrarily large region as time tends to infinity no matter how weak the noise is.

Proof of Theorem 7. 1 At $t = 0$, the opinions form N_c separated static clusters as $V_1, \cdots, V_{Nc}$, and the noises only affect agent $1 \in V_1$ whose opinion value is at the bottom in the beginning. Now let us consider clusters V_1 and V_2.

By Lemma 7.2, the opinions of V_1 driven by noises will finally interact with opinions of cluster V_2 at some time, and before that moment, agents of cluster V_1 possess the same opinion value except agent 1, whose opinion value has an additive $\xi(t)$ according to (7.4). This means that it is agent 1 of V_1 who will firstly interact with the agents of V_2. That is, a. s. there is a moment $T_1 > 0$ such that for all $i \in V_2$,

$$x_i(T_1) - x_1(T_1) \leqslant \varepsilon$$

Denote

$$\overline{T} = \inf\{t:\ x_i(t) - x_1(t) \leqslant \varepsilon,\ i \in V_2\} \tag{7.8}$$

Then $\mathbb{P}\{\overline{T} < \infty\} = 1$ and a. s.

$$\begin{aligned} &x_i(\overline{T}) - x_1(\overline{T}) \leqslant \varepsilon \\ &x_i(\overline{T}) - x_j(\overline{T}) > \varepsilon,\ j \in V_1 - \{1\} \end{aligned} \tag{7.9}$$

as well as a. s.

$$d_{V_1}(\overline{T}) \leqslant \delta_2 \tag{7.10}$$

Then by (7.4)

$$\begin{aligned} x_i(\overline{T}+1) &= \frac{1}{1+|V_2|}\Big(x_1(\overline{T}) + \sum_{k \in V_2} x_k(\overline{T})\Big) \\ &= \frac{1}{1+|V_2|}x_1(\overline{T}) + \frac{|V_2|}{1+|V_2|}x_i(\overline{T}),\ i \in V_2 \end{aligned} \tag{7.11}$$

and

$$\begin{aligned} x_j(\overline{T}+1) &= \frac{1}{|V_1|}\Big(x_1(\overline{T}) + \sum_{k \in V_1 - \{1\}} x_k(\overline{T})\Big) \\ &= \frac{1}{|V_1|}x_1(\overline{T}) + \frac{|V_1|-1}{|V_1|}x_j(\overline{T}) \\ &= \frac{1}{|V_1|}x_1(\overline{T}) + \frac{|V_1|-1}{|V_1|}(x_1(\overline{T}) - \xi(\overline{T})) \\ &= x_1(\overline{T}) - \frac{|V_1|-1}{|V_1|}\xi(\overline{T}),\ j \in V_1 - \{1\} \end{aligned} \tag{7.12}$$

and

$$\begin{aligned} x_1(\overline{T}+1) &= \frac{1}{|V_1|+|V_2|}\Big(\sum_{j \in V_1} x_j(\overline{T}) + \sum_{i \in V_2} x_2(\overline{T})\Big) + \xi(\overline{T}+1) \\ &= \frac{1}{|V_1|+|V_2|}(|V_1|x_1(\overline{T}) + |V_2|x_i(\overline{T})) - \frac{|V_1|-1}{|V_1|+|V_2|}\xi(\overline{T}) + \xi(\overline{T}+1) \\ &\leqslant \frac{1}{1+|V_2|}(x_1(\overline{T}) + |V_2|x_i(\overline{T})) - \frac{|V_1|-1}{|V_1|+|V_2|}\xi(\overline{T}) + \xi(\overline{T}+1) \end{aligned}$$

$$\leqslant x_1(\bar{T}) + \frac{|V_2|}{1+|V_2|}\varepsilon - \frac{|V_1|-1}{|V_1|+|V_2|}\xi(\bar{T}) + \xi(\bar{T}+1),\ i \in V_2 \quad (7.13)$$

where the last two inequalities follow from $|V_1| \geqslant 1$, (7.9) and Lemma 3.1. By (7.11) and (7.12), we have a.s.

$$\begin{aligned} x_i(\bar{T}+1) - x_j(\bar{T}+1) &= \frac{|V_2|}{1+|V_2|}(x_i(\bar{T}) - x_1(\bar{T})) + \frac{|V_1|-1}{|V_1|}\xi(\bar{T}) \\ &\leqslant \frac{|V_2|}{1+|V_2|}\varepsilon + \frac{|V_1|-1}{|V_1|}\delta_2,\ i \in V_2,\ j \in V_1 - \{1\} \end{aligned} \quad (7.14)$$

and by (7.12), (7.13), we have a.s.

$$\begin{aligned} x_1(\bar{T}+1) - x_j(\bar{T}+1) &\leqslant \frac{|V_1|-1}{|V_1|}\xi(\bar{T}) + \frac{|V_2|}{1+|V_2|}\varepsilon - \frac{|V_1|-1}{|V_1|+|V_2|}\xi(\bar{T}) + \xi(\bar{T}+1) \\ &\leqslant \frac{n-1}{n}\varepsilon + \left(1 - \frac{1}{n}\right)\delta_2 + \delta_2 = \frac{n-1}{n}\varepsilon + \left(2 - \frac{1}{n}\right)\delta_2 \end{aligned} \quad (7.15)$$

since $\frac{1}{|V_1|} + \frac{|V_1|-1}{|V_1|+|V_2|} \geqslant \frac{1}{n}$ holds for $1 \leqslant |V_1|$, $|V_2| \leqslant n$, $|V_1| + |V_2| \leqslant n$. If $\delta_2 \leqslant \frac{\varepsilon}{2n}$, by (7.14) and (7.15), we have a.s.

$$\begin{aligned} x_i(\bar{T}+1) - x_j(\bar{T}+1) &\leqslant \varepsilon \\ x_1(\bar{T}+1) - x_j(\bar{T}+1) &\leqslant \varepsilon \end{aligned} \quad (7.16)$$

(7.16) implies that at $\bar{T}+1$, all agents of V_1 and V_2 are neighbors to each other, then by Lemma 7.1, we have $d_{V_1 \cup V_2}(t) \leqslant \delta_2$, $t \geqslant \bar{T}+1$. Repeating the above procedure with respect to clusters of V_3 to V_{Nc}, we will have $d_V(t) \leqslant \delta_2$, $t \geqslant \bar{T}^*$ with $\mathbb{P}\{\bar{T}^* < \infty\} = 1$. This completes the proof. □

7.3 Soft Control Strategy

In Section 7.2, we show that weak noise can eliminate the disagreement of opinions and induce the consensus of system. However, Lemma 7.2 shows that the synchronized opinions driven by random noise could go to infinity, which contradicts with our practical wish to control opinions to some objective value. In this part, we will design a simple soft control strategy to guide the opinions to the targeted opinion value. The so-called soft control means that one does not control the agents directly but put additional agents (called shill) in the group to intervene the behavior of original agents. In this soft control strategy, we introduce in the system a "leader" agent who is closed-minded and sticks to the fixed opinion value, but other agents will take into account the leader's opinion once the leader locates in their neighbor region. Then we prove that the opinions will almost surely gather to the leader's opinion in finite time.

To demonstrate the sketch of the above soft control scheme, denote the leader as agent l, and we simply set l's opinion value to be

$$x_l(t) \equiv A > x_{Nc}^* + \varepsilon,\ t \geqslant 0 \tag{7.17}$$

in the divisive system (7.2) ~ (7.5), where the neighbor set in (7.2) is modified as

$$N_i(x(t)) = \{j \in V \cup \{l\} \mid |x_j(t) - x_i(t)| \leqslant \varepsilon\},\ i \in V$$

then we prove

Theorem 7.2 Consider the divisive system (7.2) ~ (7.5) with an additional leader agent l whose opinion value satisfies (7.17) and denote $d_V^A(t) = \max_{i \in V} |x_i(t) - A|$. Then still with the conditions given in Theorem 7.1, we have a. s. $\limsup_{t \to \infty} d_V^A(t) \leqslant 2\delta_2$.

Remark 7.1 Here we only demonstrate a simple scheme by setting the leader above all the agents for convenience. If we consider the case when the leader agent can be placed anywhere, such as in the intermediate region of the divisive opinions, the disagreement elimination strategy can still work by flexibly introducing noise to more agents, for instance, two agents in distinct subgroups.

Proof of Theorem 7.2 Theorem 7.1 tells that the opinions of all agents in V will reach δ_2-consensus in finite time T, and suppose $T = 0$ a. s. without loss of generality.

If $x_i(t) < A - \varepsilon,\ i \in V$, it is easy to get that a. s.

$$|x_1(t+1) - x_j(t)| \leqslant \delta_2 \leqslant \varepsilon,\ j \in V - \{1\} \tag{7.18}$$

If $\max\{d_V(t),\ d_V^A(t)\} \leqslant \varepsilon$ a. s., we have for $i \in V$

$$\begin{aligned} x_i(t+1) &= \frac{1}{n+1}\left(A + \sum_{k \in V} x_k(t)\right) + I_{\{i=1\}}\xi(t+1) \\ &= \frac{1}{n+1}A + \frac{n}{n+1}x_j(t) + \frac{1}{n+1}\xi(t) + I_{\{i=1\}}\xi(t+1),\ j \in V - \{1\} \end{aligned} \tag{7.19}$$

then a. s.

$$d_V(t+1) \leqslant |\xi(t+1)| \leqslant \delta_2 \leqslant \varepsilon$$

$$|A - x_i(t+1)| \leqslant \left|\frac{n}{n+1}(A - x_i(t))\right| + \frac{|\xi(t)|}{n+1} + |\xi(t+1)|$$

$$\leqslant \frac{n}{n+1}\varepsilon + \frac{n+2}{n+1}\delta_2 \leqslant \varepsilon, \ i \in V \tag{7.20}$$

since $\delta_2 \leqslant \frac{\varepsilon}{2n}$. Repeating the above procedure, it obtains that once (7.20) holds at some moment, it holds forever.

Next we will show that $d_V(t) \leqslant \varepsilon$, $t \geqslant 0$ a. s. $d_V(t) \leqslant \varepsilon$, $t \geqslant 0$ a. s.. If not, by (7.18) and (7.20), it is easy to know that it could only happen when $x_1(t) \in [A-\varepsilon, A)$ and $x_j(t) \in [A-\varepsilon-\delta_2, A-\varepsilon]$. In this case

$$\begin{aligned} x_1(t+1) &= \frac{1}{n+1}\Big(A + x_1(t) + \sum_{j \in V-\{1\}} x_j(t)\Big) + \xi(t+1) \\ &= \frac{A + x_1(t)}{n+1} + \frac{n-1}{n+1}x_j(t) + \xi(t+1) \\ x_j(t+1) &= \frac{1}{n}\Big(x_1(t) + \sum_{k \in V-\{1\}} x_k(t)\Big) \\ &= \frac{n-1}{n}x_j(t) + \frac{1}{n}x_1(t), \ j \in V - \{1\} \end{aligned}$$

yielding

$$\begin{aligned} x_1(t+1) - x_j(t+1) &= \frac{A - x_j(t)}{n+1} - \frac{1}{n(n+1)}(x_1(t) - x_j(t)) + \xi(t+1) \\ &\leqslant \frac{\varepsilon + \delta_2}{n+1} + \delta_2 \leqslant \varepsilon, \text{ a. s.} \end{aligned} \tag{7.21}$$

Hence

$$d_V(t) \leqslant \varepsilon, \ t \geqslant 0 \quad \text{a. s.} \tag{7.22}$$

Denote the original opinion values that satisfy (7.2) ~ (7.5) without introducing the leader agent as $y_i(t)$, $i \in V$, $t \geqslant 1$, and

$$T_l = \inf\{t: d_V^A(t) \leqslant \varepsilon\} \tag{7.23}$$

Then by Lemma 3.1 and (7.22), we have

$$x_i(t) \geqslant y_i(t) \quad \text{a. s.}, \ t \leqslant T_l \tag{7.24}$$

By Lemma 7.2, we have

$$\mathbb{P}\{1 \leqslant T_l < \infty\} = 1 \tag{7.25}$$

By (7.20), (7.22) and (7.25)

$$\max\{d_V(t), d_V^A(t)\} \leqslant \varepsilon, \text{ a. s. } t \geqslant T_l \tag{7.26}$$

Then on $\{1 \leqslant T_l < \infty\}$, for $j \in V - \{1\}$, $t \geqslant T_l$, by (7.19)

$$\begin{aligned}
|A - x_j(t+1)| &\leqslant \frac{n}{n+1}|A - x_j(t)| + \frac{1}{n+1}|\xi(t)| \\
&\leqslant \left(\frac{n}{n+1}\right)^2 |A - x_j(t-1)| + \frac{n}{n+1}\frac{|\xi(t-1)|}{n+1} + \frac{|\xi(t)|}{n+1} \\
&\vdots \\
&\leqslant \left(\frac{n}{n+1}\right)^{t-T_l+1} |A - x_j(T_l)| + \sum_{k=T_l}^{t}\left(\frac{n}{n+1}\right)^{t-k}\frac{|\xi(k)|}{n+1} \\
&\leqslant \left(\frac{n}{n+1}\right)^{t-T_l+1} |A - x_j(T_l)| + (n+1)\left(1 - \left(\frac{n}{n+1}\right)^{t-T_l+1}\right)\frac{\delta_2}{n+1}, \text{ a. s.}
\end{aligned} \tag{7.27}$$

whence

$$\limsup_{t\to\infty}|A - x_j(t+1)| \leqslant \delta_2, \text{ a. s. }, j \in V - \{1\} \tag{7.28}$$

and a. s.

$$\limsup_{t\to\infty}|A - x_1(t+1)| \leqslant \limsup_{t\to\infty}(|A - x_j(t+1)| + |x_1(t+1) - x_j(t+1)|) \leqslant 2\delta_2 \tag{7.29}$$

yielding the conclusion. □

7.4 Stopping Time of Reaching ϕ-Consensus

Theorem 7.1 tells that the divisive system (7.2) ~ (7.5) can a. s. achieve δ_2-consensus in finite time under the drive of weak random noise, and Theorem 7.2 shows that the opinions will run to the objective value in finite time with the soft control of a leader agent. Then a natural question arising subsequently is what the finite time is. Actually, by the proof of Theorem 7.1, the time T when the noisy system (7.2) ~ (7.5) reach δ_2-consensus is a stopping time which is finite with $\mathbb{P}\{T = \infty\} = 0$. For a finite stopping time, its expectation could be finite or infinite, then with the conditions given in Theorem 7.1, we have

Theorem 7.3 Let $T = \inf\{t: d_V(t) \leqslant \varepsilon\}$, then

(1) $\mathbf{E}T=\infty$, if $\delta_1=\delta_2$.

(2) $\mathbf{E}T\leqslant\frac{2n}{\delta_2-\delta_1}(x^*_{N_c}-x^*_1+(N_c-1)\delta_2)$, if $\delta_1<\delta_2$.

Remark 7.2 Usually, we say a finite stopping time T is integrable if $\mathbf{E}T<\infty$, then Theorem 7.3 implies that the finite stopping time when the opinion differences are eliminated is integrable only when the noise is oriented. Also, it can be seen from the coefficient $\frac{2n}{\delta_2-\delta_1}$ that given $0<\delta_2\leqslant\frac{\varepsilon}{2n}$, a tiny increase of δ_1 could dramatically decrease the mean of the stopping time, which will be illustrated by the simulations in next section. Actually, the formula for the integrable stopping time in oriented case integrates all elements that intuitively involve with the stopping time in divisive system (7.2) ~ (7.5).

To prove Theorem 7.3, some preliminary lemmas are needed.

Lemma 7.3 Suppose $\{X_t,\ t\geqslant1\}$ are i. i. d. random variables with $\mathbf{E}X_1=\mu\geqslant0$, $\mathbf{E}|X_1|>0$, and let $S_t=\sum_1^t X_i$, $T_0=\inf\{t\geqslant1:\ S_t\geqslant0\}$, $T_c=\inf\{t\geqslant1:\ S_t>c>0\}$, then $\mathbf{E}T_c\geqslant\mathbf{E}T_0=\infty$ for $\mu=0$ and $\mathbf{E}T_c<\infty$ for $\mu>0$.

Proof Consider Exercise 5.4.11. of [9] (Ex. 11), then when $\mu=0$, we have $\mathbf{E}T_c=\infty$, whence $\mathbf{E}T_0=\infty$ by Theorem 5.4.1 of [9]. Apropos of the case of $\mu>0$, denote $T_0^+=\inf\{t\geqslant1:\ S_t>0\}$, then $\mathbf{E}T_0^+<\infty$ by Ex. 11, and hence $\mathbf{E}T_0\leqslant\mathbf{E}T_0^+<\infty$, which implies $\mathbf{E}T_c<\infty$ by Theorem 5.4.1 of [9]. □

Lemma 7.4 Suppose $\{X_t,\ t\geqslant1\}$ are i. i. d random variables uniformly distributed on $[-\delta,\delta]$ with $\delta>0$ and denote $Q_1=X_1$, $Q_t=\sum_1^{t-1}X_i+\alpha X_t$, $t\geqslant2$ with $\alpha>1$. Let $T_0'=\inf\{t\geqslant1:\ Q_t\geqslant0\}$, then $ET_0'=\infty$.

Proof Let $L=\left\lceil\frac{(\alpha-1)\delta}{a}\right\rceil$, $S_t=\sum_1^t X_i$, $U_t^k=\sum_{t+1}^{t+k}X_i$. For $t>L$, if $X_1<-a,\cdots,X_L<-a$, $\max_{1\leqslant k\leqslant t-L}U_L^k<0$, we can get for all $1\leqslant j\leqslant t$ that

$$Q_j\leqslant S_j+(\alpha-1)X_j=S_L+U_L^{j-L}+(\alpha-1)X_j\leqslant -aL+U_L^{j-L}+(\alpha-1)\delta<0$$

implying

$$\left\{X_1<-a,\cdots,X_L<-a,\ \max_{1\leqslant k\leqslant t-L}U_L^k<0\right\}\subset\left\{\max_{1\leqslant j\leqslant t}Q_j<0\right\} \tag{7.30}$$

By independence and Markov property

$$\begin{aligned}\mathbb{P}\left\{X_1<-a,\cdots,X_L<-a,\ \max_{1\leqslant k\leqslant t-L}U_L^k<0\right\}&=p^L\mathbb{P}\left\{\max_{1\leqslant k\leqslant t-L}U_L^k<0\right\}\\&=p^L\mathbb{P}\left\{\max_{1\leqslant k\leqslant t-L}S_k<0\right\}\end{aligned} \tag{7.31}$$

Then by (7.30) and (7.31),

$$p^L \mathbb{P}\left\{\max_{1\leqslant k\leqslant t-L} S_k < 0\right\} \leqslant \mathbb{P}\left\{\max_{1\leqslant k\leqslant t-L} Q_j < 0\right\} \tag{7.32}$$

Denote $T_0 = \inf\{t\geqslant 1: S_t \geqslant 0\}$, by Lemma 7.3, we have

$$\mathbf{E}T_0 = \infty \tag{7.33}$$

Hence by (7.32) and (7.33),

$$\begin{aligned}
\mathbf{E}T'_0 &= \sum_{t=1}^{\infty} \mathbb{P}\{T'_0 \geqslant t\} = 1 + \sum_{t=1}^{\infty} \mathbb{P}\left\{\max_{1\leqslant j\leqslant t} Q_j < 0\right\} \\
&= 1 + \sum_{t=1}^{L} \mathbb{P}\left\{\max_{1\leqslant j\leqslant t} Q_j < 0\right\} + \sum_{t=L+1}^{\infty} \mathbb{P}\left\{\max_{1\leqslant j\leqslant t} Q_j < 0\right\} \\
&\geqslant 1 + p^L \sum_{t=L+1}^{\infty} \mathbb{P}\left\{\max_{1\leqslant j\leqslant t} S_k < 0\right\} \\
&= 1 + p^L \sum_{t=1}^{\infty} \mathbb{P}\left\{\max_{1\leqslant j\leqslant t} S_k < 0\right\} \\
&= 1 + p^L \sum_{t=1}^{\infty} \mathbb{P}\{T_0 \geqslant t+1\} \\
&= 1 + p^L(ET_0 - 1) = \infty
\end{aligned}$$

This completes proof. □

In Lemma 7.4, we define a new type of stopping time. Compared to the classical stopping time T_0 defined in Lemma 7.3, it is easy to check that T'_0 gets shorter. However, this variation is not essential, since Lemma 7.4 shows that T'_0 is still not integrable.

Lemma 7.5[9] (Wald's Equation). Let $\{X_t, t\geqslant 1\}$ are i. i. d. random variables, and $S_t = \sum_{1}^{t} X_i$, $t\geqslant 1$. If $\mathbf{E}X_1$ exists and T is an $\{X_n\}$ – stopping time with $\mathbf{E}T < \infty$, then

$$\mathbf{E}S_T = \mathbf{E}X_1 \cdot \mathbf{E}T$$

Proof of Theorem 7.3 Considering the stopping time $\overline{T}$ defined in (7.8), it is known from the proof of Theorem 7.1 that at time $\overline{T}+1$, clusters V_1 and V_2 achieve δ_2-consensus. Hence, first we need to calculate the stopping time $\overline{T}$.

By (7.8) and (7.9), we know that at time $\overline{T}$, only agent 1 of group V_1 becomes the neighbor of agents in group V_2. Then for $0 < t \leqslant \overline{T}$,

$$x_1(t) = \frac{1}{|V_1|}\sum_{j\in V_1} x_j(t-1) + \xi(t)$$

$$
\begin{aligned}
&= \frac{1}{|V_1|}\Big(\sum_{j\in V_1}\frac{1}{|V_1|}\sum_{k\in V_1}x_k(t-2) + \xi(t-1)\Big) + \xi(t) \\
&= \frac{1}{|V_1|}\sum_{j\in V_1}x_j(t-2) + \frac{1}{|V_1|}\xi(t-1) + \xi(t) \\
&= \frac{1}{|V_1|}\sum_{j\in V_1}x_j(0) + \sum_{k=1}^{t-1}\frac{1}{|V_1|}\xi(k) + \xi(t) \\
&= x_1^* + \frac{1}{|V_1|}\sum_{k=1}^{t-1}\xi(k) + \xi(t)
\end{aligned} \tag{7.34}
$$

Let $R_1 = \xi(1)$, $R_t = \sum_{k=1}^{t-1}\xi(k) + |V_1|\xi(t)$, $t \geqslant 2$, then by (7.8), (7.9) and (7.34)

$$\overline{T} = \inf\{t>0:\ R_t \geqslant |V_1|(x_2^* - x_1^* - \varepsilon)\} \tag{7.35}$$

(1) $\delta_1 = \delta_2$: If $|V_1| = 1$, by (7.35) and Lemma 7.3, it can be obtained that $\mathbf{E}\overline{T} = \infty$. Otherwise $|V_1| > 1$, then $\mathbf{E}\overline{T} = \infty$ can be obtained by (7.35) and Lemma 7.4. Hence

$$\mathbf{E}T \geqslant \mathbf{E}\overline{T} = \infty \tag{7.36}$$

(2) $\delta_1 < \delta_2$: Let $M = |V|(x_2^* - x_1^* - \varepsilon)$, $U_t = \sum_{k=1}^{t}\xi(k)$, $t > 0$ and denote

$$T_1 = \inf\{t>0:\ U_t \geqslant M + (|V|-1)\delta_2\}$$

Since $\{R_t < M\} \subset \{U_t < M + (|V|-1)\delta_2\}$, we have

$$
\begin{aligned}
\mathbb{P}\{\overline{T} \geqslant t+1\} &= \mathbb{P}\Big\{\max_{j\leqslant t} R_j < M\Big\} \\
&\leqslant \mathbb{P}\Big\{\max_{j\leqslant t} U_j < M + (|V|-1)\delta_2\Big\} \\
&= \mathbb{P}\{T_1 \geqslant t+1\},\ t \geqslant 1
\end{aligned} \tag{7.37}
$$

Note that $\mathbf{E}\xi(1) = \frac{\delta_2 - \delta_1}{2} > 0$, then by Lemma 7.3, $\mathbf{E}T_1 < \infty$. Thus by (7.37)

$$\mathbf{E}\overline{T} = \sum_{t=1}^{\infty}\mathbb{P}\{\overline{T} \geqslant t\} \leqslant \sum_{t=1}^{\infty}\mathbb{P}\{T_1 \geqslant t\} = \mathbf{E}T_1 < \infty \tag{7.38}$$

Since $U_{T_1-1} < M + (|V|-1)\delta_2$, $\xi(1) \leqslant \delta_2$, it yields

$$U_{T_1} < M + |V|\delta_2,\ \text{a. s.} \tag{7.39}$$

By Lemma 7.5 and (7.39)

$$\mathbf{E}T_1=\frac{\mathbf{E}U_{T_1}}{\mathbf{E}\xi(1)}\leqslant\frac{2|V|(x_2^*-x_1^*-\varepsilon+\delta_2)}{\delta_2-\delta_1}\tag{7.40}$$

Hence (7.38) yields

$$\mathbf{E}\overline{T}\leqslant\frac{2|V|(x_2^*-x_1^*-\varepsilon+\delta_2)}{\delta_2-\delta_1}\tag{7.41}$$

Since $\overline{T}$ is the time when clusters V_1 and V_2 get merged, by (7.11) for $i\in V_2$

$$x_2^*-x_i(\overline{T}+1)=\frac{1}{1+|V_2|}(x_2^*-x_1(\overline{T}))\leqslant\varepsilon,\ \text{a. s.}$$

implying

$$x_3^*-x_i(\overline{T}+1)-\varepsilon\leqslant x_3^*-x_2^*,\ \text{a. s.}\tag{7.42}$$

Define $T^{(1)}=T_1$, $T^{(j+1)}=\inf\{t\geqslant 1:\ U_{T_{j+1}}-U_{T_j}\geqslant M_{j+1}+(|V|-1)\delta_2+\varepsilon\}$, $j\geqslant 1$ where $T_0=0$, $T_j=\sum_{i=1}^{j}T^{(i)}$ and $M_{j+1}=|V|(x_{j+1}^*-x_j^*)$, then $\{T^{(j)},j\geqslant 1\}$ are independent stopping variables. Following the above procedure of deriving (7.41) and considering (7.42), it obtains the time T when all clusters achieve δ_2-consensus satisfies

$$\begin{aligned}\mathbf{E}T&\leqslant\frac{2|V|}{\delta_2-\delta_1}\sum_{j=1}^{N_c-1}(x_{j+1}^*-x_j^*+\delta_2)\\&=\frac{2n}{\delta_2-\delta_1}(x_{N_c}^*-x_1^*+(N_c-1)\delta_2)\end{aligned}\tag{7.43}$$

This completes the proof. □

For the stopping time when opinions are synchronized to the leader's opinion value, we can take it to be T_l by considering (7.23) and (7.26), though (7.28) and (7.29) show that the synchronization could happen asymptotically. It is known from (7.25) that T_l is also a finite stopping time. Compared to the stopping time T when all opinions reach δ_2-consensus, T_l increases by the time amount of reaching to A for the synchronized opinions. Then by (7.36) and (7.43), we have

Theorem 7.4 Let $T_l=\inf\{t:\ d_V^A(t)\leqslant\varepsilon\}$ with $d_V^A(t)=\max_{i\in V}|x_i(t)-A|$, then

(1) $\mathbf{E}T_l=\infty$, if $\delta_2=\delta_1$;

(2) $\mathbf{E}T_l\leqslant\frac{2n}{\delta_2-\delta_1}(A-x_1^*+\varepsilon+N_c\delta_2)$, if $\delta_2<\delta_1$.

7.5 Simulations

In this part, we will present some simulation results to illustrate the main theorems. Here, we suppose the agent number is 10, then by Theorem 7.1, it is obtained that system (7.2) ~ (7.5) will reach δ_2-consensus in finite time when $0 < \delta_2 \leqslant 0.05\varepsilon$. And by Theorem 7.3, we know that the stopping time of reaching δ_2-consensus is not integrable when $0 < \delta_1 = \delta_2$, and integrable when $\delta_1 < \delta_2$. Fig. 7.1 and Fig. 7.2 illustrate that system (7.2) ~ (7.5) achieve δ_2-consensus in finite time when noises are either neutral or oriented, but the stopping time of reaching δ_2-consensus in neutral case (Fig. 7.1) is much longer than that in oriented case (Fig. 7.2). Moreover, a large number of simulations show that a tiny increase of δ_1 will sharply decrease the time when the system reach δ_2-consensus, and this can be also predicted from Theorem 7.3.

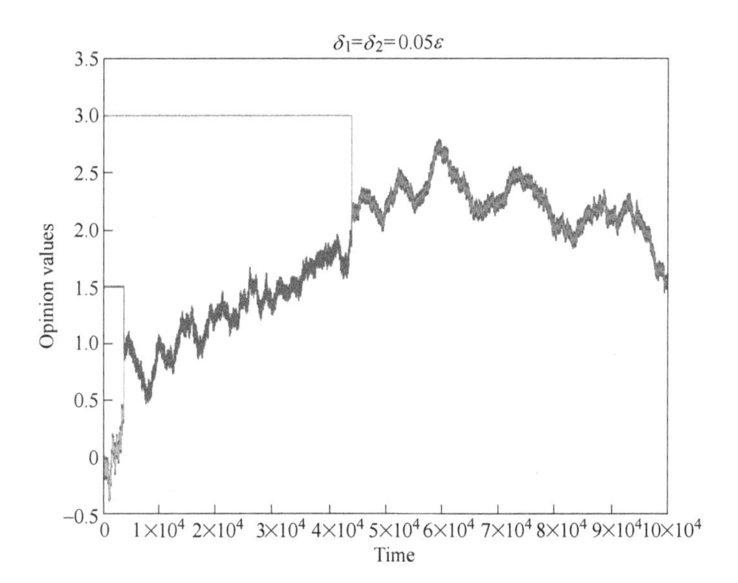

Fig. 7.1 Opinion evolution of system (7.2) ~ (7.5) of 10 agents with neutral noise uniformly distributed on $[-0.05, 0.05]\varepsilon$

(The initial opinion value $x(0) = [0\ \ 0\ \ 0\ \ 0\ \ 1.5\ \ 1.5\ \ 1.5\ \ 1.5\ \ 3\ \ 3]$, confidence threshold $\varepsilon = 1$, noise strength $\delta_1 = \delta_2 = 0.05$)

Fig. 7.3 and Fig. 7.4 illustrate the results when a leader agent is added to the system, and it can be seen that the opinions are synchronized to leader's opinion in finite time. Also, the time of gathering to leader's opinion when noise is oriented is shown to be much shorter than the neutral case even though δ_1 changes very little.

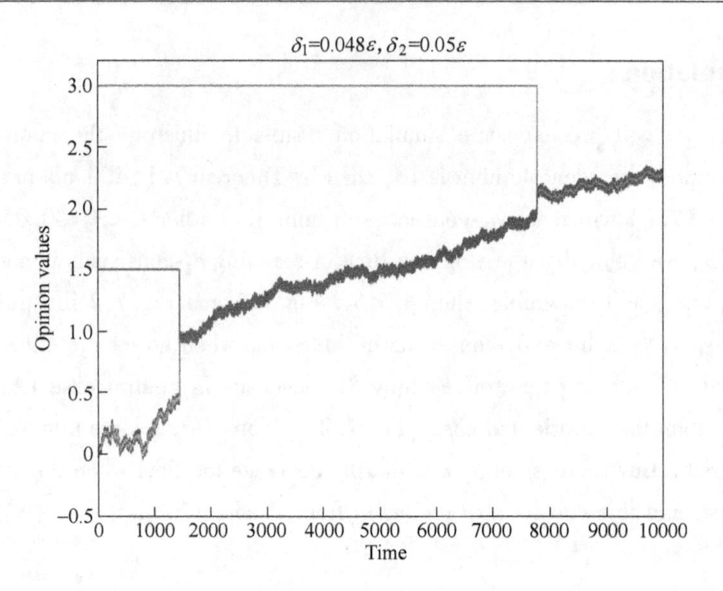

Fig. 7.2 Opinion evolution of system (7.2) ~ (7.5) of 10 agents with oriented noise uniformly distributed on [−0.048, 0.05]ε

(The initial opinion value $x(0)$ = [0 0 0 0 1.5 1.5 1.5 1.5 3 3], confidence threshold $\varepsilon = 1$, noise strength $\delta_1 = 0.048$, $\delta_2 = 0.05$)

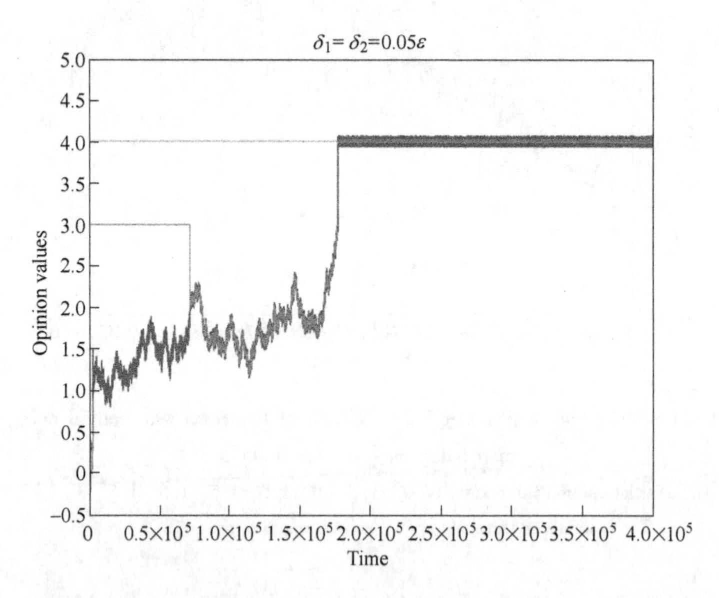

Fig. 7.3 Opinion evolution of the softly controlled system (7.2) ~ (7.5) with neutral noise uniformly distributed on [−0.05, 0.05]ε

(The initial opinion value $x(0)$ = [0 0 0 0 1.5 1.5 1.5 1.5 3 3], leader opinion value $A = 4.01$, confidence threshold $\varepsilon = 1$, noise strength $\delta_1 = \delta_2 = 0.05$)

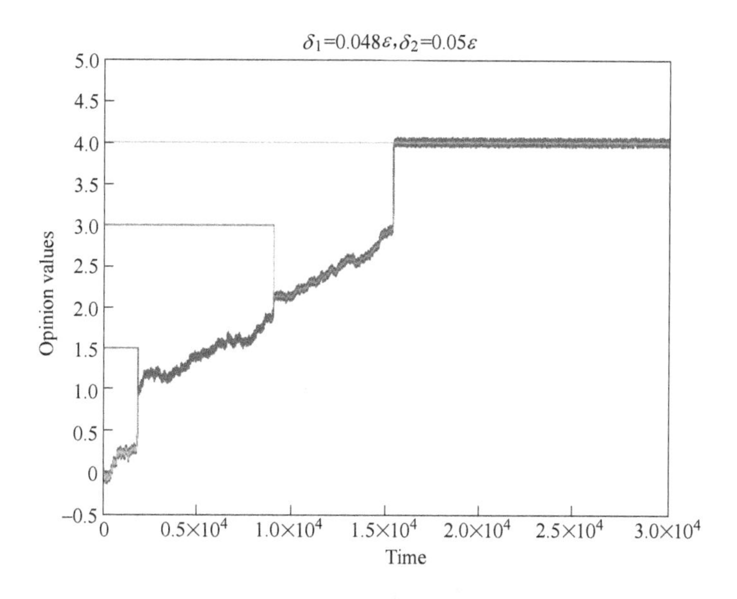

Fig. 7.4 Opinion evolution of the softly controlled system (7.2) - (7.5) with oriented noise uniformly distributed on $[-0.048, 0.05]\varepsilon$

(The initial opinion value $x(0) = [0 \ \ 0 \ \ 0 \ \ 0 \ \ 1.5 \ \ 1.5 \ \ 1.5 \ \ 1.5 \ \ 3 \ \ 3]$, leader opinion value $A = 4.01$, confidence threshold $\varepsilon = 1$, noise strength $\delta_1 = 0.048$, $\delta_2 = 0.05$)

7.6 Notes

This chapter started the design of practical noise intervention of opinion dynamics by proposing a simple intervention strategy to eliminate disagreement of a divisive system and further induce the opinions synchronized to an objective value. By introducing covert weak noise to one targeted individual and a leader agent to the system, it is firstly proved that the proposed measure can almost surely synchronize the divisive opinions to the leader's opinion in finite time. Then it is approximately calculated the finite stopping time when the noisy opinions achieve ϕ-consensus. The finite stopping time is shown not integrable when noise is neutral, and integrable when noise is oriented. This fact will provide us with further insights for designing more effective strategies to eliminate social disagreement in practice. What is more, some further noise-based control strategy of opinion dynamics can be explored in future, including the fixed-time and optimal-time noise intervention, etc.

At last, in this chapter the divisive opinion system is generated by HK dynamics, which is a self-organized system based on local rule. There are also other divisive systems generated by the topology-dependent system, such as DeGroot model. We want to mention that, the noise-based control strategy usually fails for topology-dependent systems. However, the noise intervention strategy works for the large social group, which is often a self-organized system.

References

[1] Nicholson M. Rationality and the Analysis of International Conflict[J]. Cambridge University Press, 1992.

[2] Rahim M. Managing Conflict in Organizations[J]. Transaction Publisher, 2001.

[3] Bernays E. Manipulating Public Opinion: The Why and The How[J]. American Journal of Sociology, 1928, 33 (6): 958 ~ 971.

[4] Norrander B, Wilcox C. Understanding Public Opinion[M]. CQ Press: Washington D. C 2rd Edition, 2002.

[5] Friedkin N. The problem of social control and coordination of complex systems in sociology: a look at the community cleavage[J]. IEEE Control Systems, 2015, 35 (3): 40 ~ 51.

[6] Han J, Li M, Guo L. Soft control on collective behavior of a group of autonomous agents by a shill agent[J]. Journal of Systems Science and Complexity, 2006, 19 (1): 54 ~ 62.

[7] Han H, Qiang C, Wang C, Han J. Intervention of DeGroot Model by Soft Control[C]. Chinese Control Conference, 2017, 30: 550 ~ 567.

[8] Blondel V D, Hendrickx J M, Tsitsiklis J N. On Krause's multi ~ agent consensus model with state-dependent connectivity[J]. IEEE Transactions on Automatic Control, 2009, 54 (11): 2586 ~ 2597.

[9] Chow Y, Teicher H. Probability Theory: Independence, Interchangeability, Martingales[M]. Springer Science & Business Media, 1997.

8 Robust Fragmentation of HK-Type Dynamics

One of predominantly important issues concerned in opinion dynamics is how to build model to produce social disagreement or community cleavage[1~4]. It is well known that the bounded confidence HK model can present obvious fragmentation. Meanwhile, G. Deffuant et al. established a similar-spirit model of bounded confidence that could also produce fragmentation[5]. After that, bounded confidence is believed as one remarkable factor to generate fragmentation of opinion dynamics. Other than bounded confidence, prejudiced and stubborn agents are also found as two factors that could take group cleavage as outcomes[6~8].

However, whether bounded confidence can produce robust clusters has been challenged a lot (see [4] or references therein). Especially, the theoretical studies in previous chapters reveal that the opinions under HK dynamics will spontaneously attain a consensus in the presence of noise, no matter whatever initial opinions and tiny noises are given. In the HK model, each agent updates its states by averaging the opinions of its confidence-dependent neighbors. This local rule of self-organization in HK dynamics provides an elementary mechanism to model the opinion evolution of large social groups and allows the fragmentation behavior to emerge. When random noise is admitted, the interaction of opinions becomes stochastic and more frequent and consequently leads to a consensus. Moreover, in Chapter 6 where a variation of HK model with homogeneous prejudiced agents was studied, a same phenomenon was verified, namely, under the drive of noise, the clusters get merged to the prejudice value. With these facts of noise-induced consensus of HK dynamics, we can conclude that the elegant HK model alone or with homogeneous prejudiced agents are inadequate to produce fragmentation in the noisy environment. Since random noise arises ubiquitously in natural and social systems[10~13], and also properly models the pervasive free information flow spreading in social network[9], how HK dynamics could produce robust fragmentation in the noisy environment requires further exploration. In [4], a robust clustering with respect to variation of model parameters is showed after assuming that agents can randomly pick any opinion in the whole opinion interval. In [11], clusters of noisy opinion are found when noise is properly large, but the clustering is not enduring since clusters are fluctuant and may get merged at some moments.

In this chapter, it is proposed and examined the noisy HK models with prejudiced and stubborn agents in both homogeneous and heterogeneous cases. Just as the HK model with homogeneous prejudiced agents uncovered in [9], it is proved that the HK model with homogeneous stubborn agents also fails to exhibit a robust fragmentation against noise, and the noisy opinions finally get synchronized to the stubborn agents. While the noisy HK model with heterogeneous stubborn agents will only partly show community cleavage for some specific initial opinions, it is finally demonstrated that the HK dynamics with heterogeneous prejudiced agents will remarkably generate a robust fragmentation in the presence of noise. Especially, we mathematically prove that the noise could regularize the group to form a bipartite cleavage in the heterogeneous model with two prejudices, even when the model produces several clusters in the noise-free situation. On the one hand, this discovery confirms a plainly intuitive idea that it is the innate difference within the community rather than the mere bounded confidence of individuals that accounts for the ubiquitous social cleavage; on the other hand, it further reveals the beneficial role free in-formation plays in the emergence of bipartite parties. Though there is still a long way to reveal completely the mechanism of social cleavage, the findings introduced in this chapter provide a substantial progress in this problem. Additionally, since the HK dynamics captures a typical self-organizing mechanism with local rule, the study on noise-induced synchronization or cluster properties of the HK-type dynamics will help investigate the noise-based control strategy of more local-rule based self-organizing systems like the particle robotics system[14].

8.1 ϕ-Consensus of HK-Type Models

In this part, we will examine the fragmentation of noisy HK model and its variations with homogeneous prejudiced and stubborn agents. The theoretical results will show that fragmentation of these models disappears in the presence of arbitrarily tiny noise. To begin with, the noisy HK models and definition of consensus in the noisy case are needed.

8.1.1 Noisy HK-Type Models

Suppose there are n agents in the group with $V=\{1,2,\cdots,n\}$, $x_i(t)\in[0,1]$, $(i\in V, t\geqslant 0)$ is the opinion value of agent i at time t, then the basic noisy HK model is:

$$x_i(t+1)=\begin{cases}1, & \hat{x}_i(t)>1\\ \hat{x}_i(t), & \hat{x}_i(t)\in[0,\ 1]\ \forall i\in V,\ t\geqslant 0\\ 0, & \hat{x}_i(t)<0\end{cases}\tag{8.1}$$

where

$$\hat{x}_i(t) = \frac{1}{|N_i(x(t))|}\sum_{j\in N_i(x(t))} x_j(t) + \xi_i(t+1) \tag{8.2}$$

and

$$N_i(x(t)) = \{j\in V: |x_j(t) - x_i(t)| \leqslant \varepsilon\} \tag{8.3}$$

Here, $\varepsilon \in (0, 1]$ is confidence bound, $\{\xi_i(t)\}_{i\in V, t>0}$ are random noises with independent and identically distribution (i. i. d.) and $\mathbf{E}\xi_1(1) = 0$, $\mathbf{E}\xi_2(1) > 0$, $|\xi_1(1)| \leqslant \delta$ a. s. for $\delta \geqslant 0$. $|\cdot|$ can be the cardinal number of a set or the absolute value of a real number accordingly.

An explanation of noise is the free information flow in social media. It is an era of social media today. Our opinions are affected not only by talking with other persons, but also by the huge amount of information we receive from social media every day. However, each day we cannot previously know what information we will receive, and whether or even how much it will influence our opinions. In this sense, it is a reasonable selection to model the information flow in social media as independent random disturbances. When using noise as an intervention tool to eliminate the disagreement of opinion system, the noise can be generated by randomly spreading information in social network.

Without noise in (8.2) (i. e. $\delta=0$), the system (8.1) ~ (8.2) degenerates to the original HK model, which converges in finite time in the sense that, there exist $T\geqslant 0$, $x_i^* \in [0,1]$, $i\in V$ such that $x_i(t) = x_i^*$, $t\geqslant T$. If $x_i^* = x_j^*$ for any $i, j\in V$ the opinions are said to reach consensus; otherwise, there must exist $i, j\in V$ such that $|x_i^* - x_j^*| > \varepsilon$ and fragmentation forms.

Other than bounded confidence, the prejudiced and stubborn agents are found to be other two remarkable causes of fragmentation. In this part, we will explore the noisy HK models with homogeneous prejudiced and stubborn agents. Here, "homogeneous" refers to that all prejudiced or stubborn agents possess an identically constant prejudice or fixed opinion.

The noisy HK model with homogeneous prejudiced agents is obtained by modifying (8.2) as follows:

$$\hat{x}_i(t) = (1-\alpha I_{\{i\in S_1\}})\sum_{j\in N(i,x(t))}\frac{x_j(t)}{|N_i(x(t))|} + \alpha I_{\{i\in S_1\}}J_1 + \xi_i(t+1) \tag{8.4}$$

where $S_1 \subset V$ is set of prejudiced agents, $J_1 \in [0,1]$ is prejudice value, and $\alpha \in (0,1]$ is the attraction strength of prejudice value. $I_{\{\cdot\}}$ is an indicator function which takes value 1 or 0 according to the condition · holds or not. As usual, without noise the model generates conspicuous fragmentation[7,9].

For the noisy HK model with homogeneous stubborn agents, we introduce additionally in model (8.1) ~ (8.2) a stubborn agent set, $\mathcal{B}_1$, whose opinion values satisfy

$$x_i(t) \equiv B_1,\ i \in \mathcal{B}_1,\ B_1 \in [0,1],\ t \geqslant 0 \tag{8.5}$$

and the neighbor set in (8.3) is modified as

$$N_i(x(t)) = \{j \in V \cup \mathcal{B}_1:\ |x_j(t) - x_i(t)| \leqslant \varepsilon\} \tag{8.6}$$

Then (8.1), (8.2) ~ (8.6) describe a noisy HK model with homogeneous stubborn agents. Likewise, when noise strength $\delta = 0$, the system degenerates to the noise-free one, which is claimed converging and also could produce fragmentation[17].

Note that in the noise-free case of model (8.4) ($\delta = 0$), if let $\alpha = 1$, the prejudiced model will completely degenerate to a deterministic model of stubborn agent. But in the noisy case, we would like to discriminate between the prejudiced and stubborn agent, in the sense that an extreme prejudiced agent (truth seeker) is like an independent thinker (such as a scholar in a frontier field), who has a tendency to explore independently but is not refusal to environmental in-formation; while a stubborn agent absolutely sticks to a fixed opinion and deny any external information (such as a broadcast which disseminates an opinion all the year round).

8.1.2 ϕ-Consensus

Due to persistent disturbance of noise, the definition of consensus in noisy HK models slightly differs and we adopt the definition of ϕ-consensus in Chapter 6:

Definition 8.1 Define

$$d_V(t) = \max_{i,j \in V} |x_i(t) - x_j(t)| \text{ and } d_V = \limsup_{t \to \infty} d_V(t)$$

For $\phi \in [0, \infty)$,

(1) if $d_V \leqslant \phi$, we say the system will reach ϕ-consensus.

(2) If $\mathbb{P}\{d_V \leqslant \phi\} = 1$, we say almost surely (a. s.) the system will reach ϕ-consensus.

(3) let $T = \inf\{t:\ d_V(t') \leqslant \phi \text{ for all } t' \geqslant t\}$. If $\mathbb{P}\{T < \infty\} = 1$, we say a. s. the system reaches ϕ-consensus in finite time.

Furthermore for a constant $A \in [0,1]$, define $d_V^A(t) = \max_{i \in V} |x_i(t) - A|$ and $d_V^A = \limsup_{t \to \infty} d_V^A(t)$, if substituting d_V^A for d_V in (1) - (3), the system is accordingly said to reach ϕ-consensus with A.

8.1.3 Spontaneous ϕ-Consensus Caused by Noise

Though HK model with either prejudiced or stubborn agents enables fragmentation, the introduction of random noise, even very tiny, makes ϕ-consensus of opinions spontaneously emerge. To be specific, we have

Theorem 8.1 Given any $x(0) \in [0,1]^n$, $\varepsilon \in (0,1]$:

(1) for all $\delta \in \left(0, \frac{\varepsilon}{2}\right]$, a. s. the system (8.1) ~ (8.2) will reach 2δ-consensus in finite time.

(2) for all $\delta \in (0, \underline{\delta}]$, a. s. the system (8.1), (8.3), (8.4) will reach $\bar{\delta}$-consensus with J_1. Here, $\bar{\delta}$ and $\underline{\delta}$ are constants determined by parameters n, ε, α, $|S_1|$.

(3) for all $\delta \in \left(0, \frac{\varepsilon}{2(n+1)}\right)$, a. s. the system (8.1), (8.2) ~ (8.6) will reach 2δ-consensus and meanwhile $(n+1)\delta$-consensus with B_1.

Theorem 8.1 evidently shows that in the presence of tiny noise (δ is tiny), the HK-based opinion dynamics with homogeneous variations lose their ability of generating fragmentation, and almost surely attain a fairly approximate consensus (see Fig. 8.2). Conclusions (1) and (2) follow directly from Theorem 3.2 of Chapter 3 and Theorem 6.1 of Chapter 6. As for conclusion (3), its proof is quite similar to that of Theorem 6.1 of Chapter 6. The main modification is the replacement of Lemma 3.2 in [9] by the following critical lemma:

Lemma 8.1 Let $0 < \delta < \frac{\varepsilon}{2(n+1)}$ and suppose there is a finite time $T \geqslant 0$ such that $d_V^{B_1}(T) \leqslant (n+1)\delta$, then $d_V \leqslant 2\delta$, $d_V^{B_1} \leqslant (n+1)\delta$.

Proof It is easy to check that $d_V(T) \leqslant 2|d_V^{B_1}| \leqslant \varepsilon$. This implies that at time T, all agents in V are neighbors to each other. By (8.1), (8.2) ~ (8.6), we have for $i \in V$

$$x_i(T+1) = \frac{\sum_{j\in V} x_j(T) + |\mathcal{B}_1| B_1}{n + |\mathcal{B}_1|} + \xi_i(T+1)$$

thus

$$|x_i(T+1) - B_1| \leqslant \left| \frac{\sum_{j\in V} x_j(T) + |\mathcal{B}_1| B_1}{n + |\mathcal{B}_1|} - B_1 \right| + |\xi_i(T+1)|$$

$$\leqslant \frac{\sum_{j\in V} |x_j(T) - B_1|}{n + |\mathcal{B}_1|} + \delta \leqslant \frac{n(n+1)}{n + |\mathcal{B}_1|}\delta + \delta$$

$$\leqslant (n+1)\delta \leqslant \frac{\varepsilon}{2}, \text{ a. s.} \tag{8.7}$$

Meanwhile, for all $i, j \in V$, we have a. s.

$$|x_i(T+1) - x_j(T+1)| \leqslant |\xi_i(T+1)| + |\xi_j(T+1)| \leqslant 2\delta$$

Repeating the above procedure yields the conclusion. □

Lemma 8.1 implies that once all the noisy opinions enter the neighbor region of stubborn agents, they will never get far away, and further reach 2δ-consensus.

In Theorem 3.1 of [9], noises are assumed to be independent and uniformly distributed on $[-\delta, \delta]$ while noise here is assumed to be i. i. d. To complete the proof of conclusion (3), we further need the following straightforward lemma:

Lemma 8.2 For i. i. d. random variables $\{\xi_i(t), i \in V, t \geqslant 1\}$ with $\mathbf{E}\xi_1(1) = 0$, $\mathbf{E}\xi_2(1) > 0$, there exist constants $a > 0$ and $0 < p \leqslant 1$, such that

$$\mathbb{P}\{\xi_i(t) > a\} \geqslant p, \ \mathbb{P}\{\xi_i(t) < -a\} \geqslant p$$

Proof of Theorem 8.1(3) By Lemma 8.2, there exist $0 < a \leqslant \delta$, $0 < p \leqslant 1$ such that for all $i \in V$, $t \geqslant 1$,

$$\mathbb{P}\{\xi_i(t) > a\} \geqslant p, \ \mathbb{P}\{\xi_i(t) < -a\} \geqslant p \tag{8.8}$$

The rest of proof quite follows the procedure after (12) in the proof of Theorem 3.1 in [9]. □

At last of this section, we give some intuitive explanations about how noise can induce consensus of HK model. Actually, HK model captures an averaging mechanism based on state-dependent local rule, which makes the system possess an attraction region of staying consensus (that is, when all the opinions are neighbors to each other, such as $\max_{i,j} |x_i - x_j| \leqslant \varepsilon$ in the original HK model). Noise can drive the opinions to finally get in the attraction region from any initial state. This coincides with the principle of "order from noise" of Heinz von Foerster[18].

8.2 HK Model with Heterogenous Prejudices

"Heterogeneous" means that the prejudiced agents in group possess distinct preferences. Heterogeneous prejudices, especially bipartite preferences exist ubiquitously in real society. Taking right-wing and left-wing in politics for example, different political inclinations produce much social cleavage. In this part, we introduce a noisy HK models with bipartite prejudices, afterwards establish theoretical results of fragmentation. The results show that clusters exist robustly in the presence of noise. Moreover, it is revealed that noise plays a significant role in regularizing the spontaneous emergence of bipartite cleavage.

The noisy model with heterogeneous prejudices is obtained by extending (8.4):

$$\hat{x}_i(t) = (1-\alpha)\sum_{j\in N_i(x(t))}\frac{x_j(t)}{|N_i(x(t))|} + \alpha(I_{\{i\in S_1\}}J_1 + I_{\{i\in S_2\}}J_2) + \xi_i(t+1) \tag{8.9}$$

where $S_1 \cup S_2 = V$, $S_1 \cap S_2 = \varnothing$ are the sets of heterogeneous prejudiced agents, and J_1, $J_2 \in [0,1]$ are the distinct prejudiced opinion values satisfying $|J_1 - J_2| > \varepsilon$.

It is easy to examine that without noise, system (8.1), (8.3), (8.9) will produce fragmentation. The following results show that cleavage is preserved in the presence of noise.

Lemma 8.3 Consider system (8.1), (8.3), (8.9), given any $x(0)\in[0,1]^n$, $\varepsilon\in(0,1)$, then (1) S_1 a. s. attains $\frac{(1-\alpha)\varepsilon+\delta}{\alpha}$-consensus with J_1; (2) S_2 a. s. attains $\frac{(1-\alpha)\varepsilon+\delta}{\alpha}$-consensus with J_2.

Proof Consider $i\in S_1$, then by (8.9), we have for $t\geqslant 0$

$$x_i(t+1) = \alpha J_1 + (1-\alpha)\sum_{j\in N_i(x(t))}\frac{x_j(t)}{|N_i(x(t))|} + \xi_i(t+1)$$

Noting that $|x_i(t) - x_j(t)| \leqslant \varepsilon$ for all $j\in N_i(x(t))$, it yields

$$\begin{aligned}
|x_i(t+1) - J_1| &= \left|(1-\alpha)\sum_{j\in N_i(x(t))}\frac{x_j(t)-J_1}{|N_i(x(t))|} + \xi_i(t+1)\right| \\
&\leqslant (1-\alpha)\sum_{j\in N_i(x(t))}\frac{|x_j(t)-J_1|}{|N_i(x(t))|} + |\xi_i(t+1)| \\
&\leqslant (1-\alpha)\sum_{j\in N_i(x(t))}\frac{|x_i(t)-J_1|+\varepsilon}{|N_i(x(t))|} + \delta \\
&= (1-\alpha)|x_i(t) - J_1| + (1-\alpha)\varepsilon + \delta \\
&\cdots \\
&\leqslant (1-\alpha)^{t+1}|x_i(0)-J_1| + ((1-\alpha)\varepsilon+\delta)((1-\alpha)^t+\cdots+(1-\alpha)+1) \\
&\to \frac{(1-\alpha)\varepsilon+\delta}{\alpha}, \text{ a. s.}, \quad \text{as } t\to\infty
\end{aligned} \tag{8.10}$$

implying $d_{S_1}^{J_1} \leqslant \frac{(1-\alpha)\varepsilon+\delta}{\alpha}$, a. s. Similarly, we can get that $d_{S_2}^{J_2} \leqslant \frac{(1-\alpha)\varepsilon+\delta}{\alpha}$, a. s.

Lemma 8.3 says that the prejudiced agents will finally stay near their preferences and the system generally displays a rough robustness of fragmentation. Smaller ε and δ, and larger α make a closer reaching to J_1 and J_2. Meanwhile, it is intuitive that for fixed ε, δ, α, a, larger discrepancy between J_1 and J_2 makes a more visible fragmentation. Though all agents in the group admit two distinct prejudices, sometimes the noise-free fragmentation forms with more than two clusters (see Fig. 8.3). However, in the presence of noise, it is found that bipartite cleavage emerges in the group and opinions of prejudiced agents go to their preferences (see Fig. 8.4). The following theorem gives a guarantee of bipartite fragmentation when the discrepancy of J_1 and J_2 is large enough.

Theorem 8.2 Suppose $J_1 - J_2 \geqslant \varepsilon + 2\frac{(1-\alpha)\varepsilon}{\alpha}$ in system (8.1), (8.3), (8.9) with $0<\alpha<1$, then for any $x(0) \in [0,1]^n$ and $\delta \in (0,(1-\alpha)\varepsilon)$, we have (1) S_1 a. s. attains $\frac{\delta}{\alpha}$-consensus with J_1; (2) S_2 a. s. attains $\frac{\delta}{\alpha}$-consensus with J_2.

Remark 8.1 Theorem 8.2 reveals that noise can regularize the prejudiced agents to their prejudices from any initial state. A natural question arising here is how much the discrepancy of J_1 and J_2 allows the "regularizing effect" of noise. Theorem 8.2 provides a lower bound. When $J_1 - J_2 \geqslant \varepsilon + 2\frac{(1-\alpha)\varepsilon}{\alpha}$, the conclusion is easily accessible; however, when $J_1 - J_2 = \varepsilon + 2\frac{(1-\alpha)\varepsilon}{\alpha}$, the conclusion is far from obvious. A further reduction of the lower bound is substantially difficult and unresolved in this chapter.

To begin with, some preliminary lemmas are need.

Lemma 8.4 Suppose $J_1 - J_2 \geqslant \varepsilon + 2\frac{\delta}{\alpha}$ in system (8.1), (8.3), (8.9). If a. s. there exists a finite time $T<\infty$ such that $d_{S_1}^{J_1}(T) \leqslant \frac{\delta}{\alpha}$ and $d_{S_2}^{J_2}(T) \leqslant \frac{\delta}{\alpha}$, then $d_{S_1}^{J_1} \leqslant \frac{\delta}{\alpha}$ and $d_{S_2}^{J_2} \leqslant \frac{\delta}{\alpha}$.

Proof Since $J_1 - J_2 \geqslant \varepsilon + 2\frac{\delta}{\alpha}$, at time T, any agent in S_1 cannot be the neighbor of agents in S_2, hence for all $i \in S_1$

$$
\begin{aligned}
|x_i(T+1) - J_1| &= \left|(1-\alpha)\sum_{j\in N_i(x(T))}\frac{x_j(T)-J_1}{|N_i(x(T))|} + \xi_i(T+1)\right| \\
&\leqslant (1-\alpha)\sum_{j\in N_i(x(T))}\frac{|x_j(T)-J_1|}{|N_i(x(T))|} + |\xi_i(T+1)|
\end{aligned}
$$

$$\leqslant (1-\alpha)\frac{\delta}{\alpha}+\delta=\frac{\delta}{\alpha}, \text{ a. s.}$$

Repeating the above procedure, it follows that $d_{S_1}^{J_1}\leqslant\frac{\delta}{\alpha}$ a. s., and similarly $d_{S_2}^{J_2}\leqslant\frac{\delta}{\alpha}$ a. s. □

Proof of Theorem 8.2 By Lemma 8.3, there a. s. exists $T<\infty$ such that for all $t\geqslant T$, it has $J_1-x_i(t)\leqslant\frac{(1-\alpha)\varepsilon+2\delta}{\alpha}$, $x_j(t)-J_2\leqslant\frac{(1-\alpha)\varepsilon+2\delta}{\alpha}$, $i\in S_1$, $j\in S_2$. Without loss of generality, we assume $T=0$ a. s. For $i\in S_1$,

$$\begin{aligned}
J_1-x_i(1)&=(1-\alpha)\sum_{j\in N_i(x(0))}\frac{J_1-x_j(0)}{|N_i(x(0))|}-\xi_i(1)\\
&\leqslant(1-\alpha)(J_1-x_i(0)+\varepsilon)-\xi_i(1)\\
&\leqslant(1-\alpha)\left(\frac{(1-\alpha)\varepsilon+2\delta}{\alpha}+\varepsilon\right)-\xi_i(1)\\
&\leqslant\frac{(1-\alpha)\varepsilon+2\delta}{\alpha}-\xi_i(1)
\end{aligned}\tag{8.11}$$

Consider the following protocol:

$$\begin{cases}\xi_i(t+1)\in(a,\delta], & \text{if } i\in S_1\\ \xi_i(t+1)\in[-\delta,-a), & \text{if } i\in S_2\end{cases}\tag{8.12}$$

where $a>0$ and the following $p>0$ is given in Lemma 8.2, then we have

$$\mathbb{P}\left\{J_1-x_i(1)<\frac{(1-\alpha)\varepsilon}{\alpha}+\frac{2\delta}{\alpha}-a\right\}\geqslant p^{|S_1|}>0,\ \forall i\in S_1$$

Let $L_0=\left[\frac{2\delta}{a\alpha}\right]$, then it is easy to check that the above procedure can be repeated no more than L_0 times when $d_{S_1}^{J_1}(L_0)<\frac{(1-\alpha)\varepsilon}{\alpha}$. By independence of $\xi_i(t)$, $i\in V$, $t\geqslant1$,

$$\mathbb{P}\left\{d_{S_1}^{J_1}(L_0)<\frac{(1-\alpha)\varepsilon}{\alpha}\right\}\geqslant p^{|S_1|L_0}\tag{8.13}$$

Similarly, we have

$$\mathbb{P}\left\{d_{S_2}^{J_2}(L_0)<\frac{(1-\alpha)\varepsilon}{\alpha}\right\}\geqslant p^{|S_2|L_0}\tag{8.14}$$

Since $J_1 - J_2 \geqslant \varepsilon + 2\dfrac{(1-\alpha)\varepsilon}{\alpha}$, we know that under the protocol 8.12, any agent in S_1 is not the neighbor to any agent in S_2 at moment L_0. Now consider the subgroup S_1 at moment L_0+1, if $d_{S_1}^{J_1}(L_0+1) \leqslant \dfrac{\delta}{\alpha}$, by Lemma 8.4 we have $d_{S_1}^{J_1} \leqslant \dfrac{\delta}{\alpha}$. Otherwise, consider the following protocol: for all $i \in S_1$, $t > L_0$,

$$\begin{aligned} &\xi_i(t+1) \in (a, \delta], && \text{if } \tilde{x}_i(t) \leqslant J_1 \\ &\xi_i(t+1) \in [-\delta, -a), && \text{if } \tilde{x}_i(t) > J_1 \end{aligned} \tag{8.15}$$

Denote $\tilde{x}_i(t) = |N_i(x(t))|^{-1} \sum_{j \in N_i(x(t))} x_j(t)$, $t \geqslant 0$, then by Lemma 3.1 and (8.1), (8.3), (8.9), it has under the protocol (8.15) that

$$\begin{aligned} d_{S_1}^{J_1}(L_0+1) &< \max_{i \in S_1}\{|x_i(L_0) - J_1| - a\} \\ &\leqslant \max_{i \in S_1}\{|x_i(L_0) - J_1| - a\} \\ &< d_{S_1}^{J_1}(L_0) - a \end{aligned} \tag{8.16}$$

By (8.8) and independence of noise, we have

$$\mathbb{P}\{d_{S_1}^{J_1}(L_0+1) < d_{S_1}^{J_1}(L_0) - a\} \geqslant p^{|S_1|} \tag{8.17}$$

If $d_{S_1}^{J_1}(L_0+1) \leqslant \dfrac{\delta}{\alpha}$ a. s., the conclusion holds by Lemma 3.1. Otherwise, let $L = \left\lceil \dfrac{1-\delta/\alpha}{a} \right\rceil$ and continue the above procedure L times. By (8.16) and independence of noises, we have

$$\begin{aligned} &\mathbb{P}\left\{d_{S_1}^{J_1}(L+L_0) \leqslant \frac{\delta}{\alpha} \,\middle|\, d_{S_1}^{J_1}(L_0) < \frac{(1-\alpha)\varepsilon}{\alpha}\right\} \\ &\geqslant \mathbb{P}\{\text{protocol (8.15) occurs } L+1 \text{ times}\} \\ &\geqslant p^{|S_1|(L+1)} > 0 \end{aligned} \tag{8.18}$$

By (8.13) and (8.18) and independence of noises, it has

$$\mathbb{P}\left\{d_{S_1}^{J_1}(L+L_0) > \frac{\delta}{\alpha}\right\} < 1 - p^{|S_1|(L_0+L+1)} \tag{8.19}$$

Let $\bar{L} = L + L_0$, $\Omega = \{\omega\}$ be the sample space of probability space$(\Omega, \mathcal{F}, \mathbb{P})$ and denote events $(m \geqslant 1)$

$$E_0 = \Omega$$
$$E_m = \left\{\omega: d_{S_1}^{J_1}(t) > \frac{\delta}{\alpha},\ (m-1)\bar{L} < t \leqslant m\bar{L}\right\} \tag{8.20}$$

By Lemma 8.4 and following the same procedure of obtaining (8.19), it has for $m \geqslant 1$

$$\begin{aligned}\mathbb{P}\left\{E_m \middle| \bigcap_{j<m} E_j\right\} &\leqslant \mathbb{P}\left\{d_{S_1}^{J_1}(m\bar{L}) > \frac{\delta}{\alpha} \middle| \bigcap_{j<m} E_j\right\} \\ &\leqslant 1 - p^{|S_1|(\bar{L}+1)} < 1\end{aligned} \tag{8.21}$$

By Lemma 8.4 it must hold $\left\{d_{S_1}^{J_1} > \frac{\delta}{\alpha}\right\} \subset \bigcap_{m \geqslant 1}\left\{d_{S_1}^{J_1}(t) > \frac{\delta}{\alpha},\ (m-1)\bar{L} < t \leqslant m\bar{L}\right\}$, subsequently by (8.21)

$$\begin{aligned}\mathbb{P}\left\{d_{S_1}^{J_1} \leqslant \frac{\delta}{\alpha}\right\} &= 1 - \mathbb{P}\left\{d_{S_1}^{J_1} > \frac{\delta}{\alpha}\right\} \\ &\geqslant 1 - \mathbb{P}\left\{\bigcap_{m \geqslant 1}\left\{d_{S_1}^{J_1} > \frac{\delta}{\alpha},\ (m-1)\bar{L} < t \leqslant m\bar{L}\right\}\right\} \\ &= 1 - \mathbb{P}\left\{\bigcap_{m \geqslant 1} E_m\right\} = 1 - \mathbb{P}\left\{\lim_{m \to \infty} \bigcap_{j=1}^{m} E_j\right\} \\ &= 1 - \lim_{m \to \infty} \mathbb{P}\left\{\bigcap_{j=1}^{m} E_j\right\} \\ &= 1 - \lim_{m \to \infty} \prod_{j=1}^{m} \mathbb{P}\left\{E_j \middle| \bigcap_{k<j} E_k\right\} \\ &\geqslant 1 - \lim_{m \to \infty}\left(1 - p^{|S_1|(\bar{L}+1)}\right)^m = 1\end{aligned}$$

Similarly, we can also obtain

$$\mathbb{P}\left\{d_{S_2}^{J_2} \leqslant \frac{\delta}{\alpha}\right\} = 1$$

This completes the proof. □

At the end of this section, we give a short discussion of the noisy HK model with heterogeneous stubborn agents. Introducing in (8.5) another stubborn agent set, B_2, and the distinct stubborn opinion values satisfy

$$x_i(t) \equiv B_1,\ x_j(t) \equiv B_2,\ i \in \mathcal{B}_1,\ j \in \mathcal{B}_2,\ t \geqslant 0 \tag{8.22}$$

with $B_1, B_2 \in [0,1]$, $\mathcal{B}_1 \cap \mathcal{B}_2 = \varnothing$. The neighbor set in (8.6) is modified accordingly as

$$N_i(x(t)) = \{j \in V \cup \mathcal{B}_1 \cup \mathcal{B}_2 : |x_j(t) - x_i(t)| \leqslant \varepsilon\} \qquad (8.23)$$

Further, it assumes that $B_2 - B_1 > \varepsilon$ which implies a pronounced difference between the stubborn agents.

Theorem 8.2 clearly shows that given any initial opinion values, the system with heterogeneous prejudices will display robust fragmentation in the presence of noise. However, for the system (8.1), (8.2), (8.22), (8.23) of heterogeneous stubborn agents, whether robust fragmentation can be formed highly depends on the location of initial opinions. Here, using conclusion (c) of Theorem 8.1, we simply give some examples without further interpretation:

Example For system (8.1), (8.2), (8.22), (8.23), (1) if $x_i(0) \in [0, B_1]$ or $[B_2, 1], i \in V$ then for all $\delta \in \left(0, \dfrac{B_2 - B_1 - \varepsilon}{n+1}\right)$, we have $d_V \leqslant 2\delta$; (2) if or, $x_i(0) \in [0, B_1]$, $i \in V_1$ and $x_j(0) \in [B_2, 1]$, $j \in V_2$ where V_1, V_2 are nonempty and $V_1 \cup V_2 = V$, $V_1 \cap V_2 = \varnothing$, then for all $\delta \in \left(0, \dfrac{B_2 - B_1 - \varepsilon}{2(n+1)}\right)$, we have $d_{V_1}^{B_1} \leqslant 2\delta$, $d_{V_2}^{B_2} \leqslant 2\delta$.

8.3 Simulations

In this part, we present some simulation results to verify our main theoretical conclusions. First, we consider the original HK model with 20 agents whose initial opinion values are randomly generated in [0,1]. Let $\varepsilon = 0.2$, $\delta = 0.1\varepsilon$, then Fig. 8.1 shows that opinions finally attain a consensus. Then we consider the modified HK model with homogeneous stubborn agents. Add a stubborn agent whose fixed opinion value takes 0.5 in the original HK model, then by letting $\delta = 0.04\varepsilon$, the Fig. 8.2 shows that opinions finally synchronize to 0.5 and the fragmentation vanishes. Some simulation results of the systems with homogeneous prejudiced agents can be found in [9], and we omit them here. For the original HK model, the noisy opinions will get synchronized but fluctuated; while for the modified HK model with homogeneous stubborn agents, the noisy opinion will synchronize with the constant opinion value of the stubborn agents.

In the following, we consider the HK system with heterogeneous prejudiced agents. Let the system consists of 20 agents, whose initial opinion values are randomly generated in [0,1], $J_1 = 0.6$, $J_2 = 0.2$, confidence threshold $\varepsilon = 0.2$. Let 10 agents own the prejudice value of J_1 and the other 10 own $J_2$and attraction strength $\alpha = 0.4$. First, we present the opinion evolution without noise, and Fig. 8.3 shows that the system forms four clusters. Then we add noise with strength $\delta = 0.05\varepsilon$ in the system, and Fig. 8.4 shows that the fragmentation emerges with two clusters which locate near the prejudices J_1 and J_2.

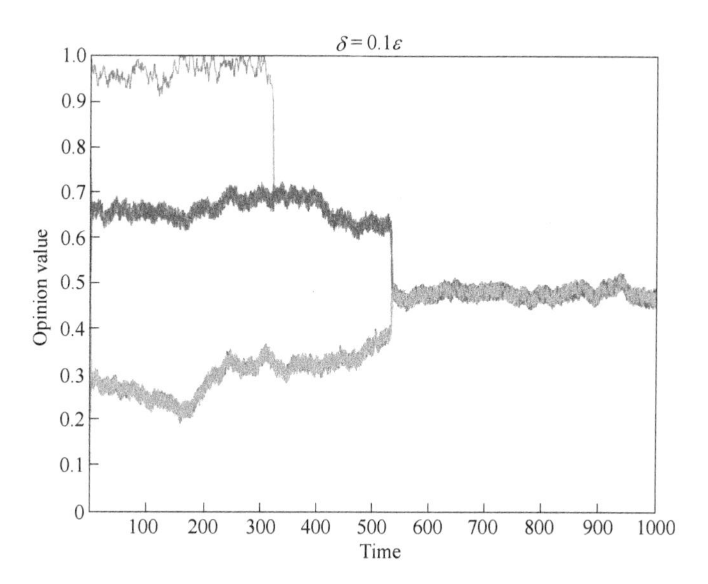

Fig. 8.1 Opinion evolution of system (8.1), (8.2) ~ (8.3) of 20 agents and a stubborn agent
(The initial opinion values are randomly generated in [0,1], confidence threshold $\varepsilon = 0.2$, and noise strength $\delta = 0.1\varepsilon$)

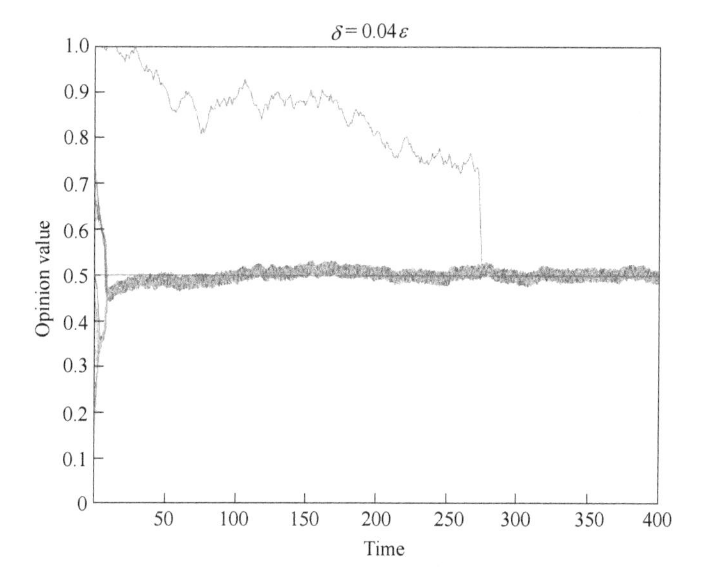

Fig. 8.2 Opinion evolution of system (8.1), (8.4) ~ (8.6) of 20 agents and a stubborn agent
(The initial opinion values are randomly generated in [0,1], the opinion value of stubborn agents takes 0.5, confidence threshold $\varepsilon = 0.2$, and noise strength $\delta = 0.04\varepsilon$)

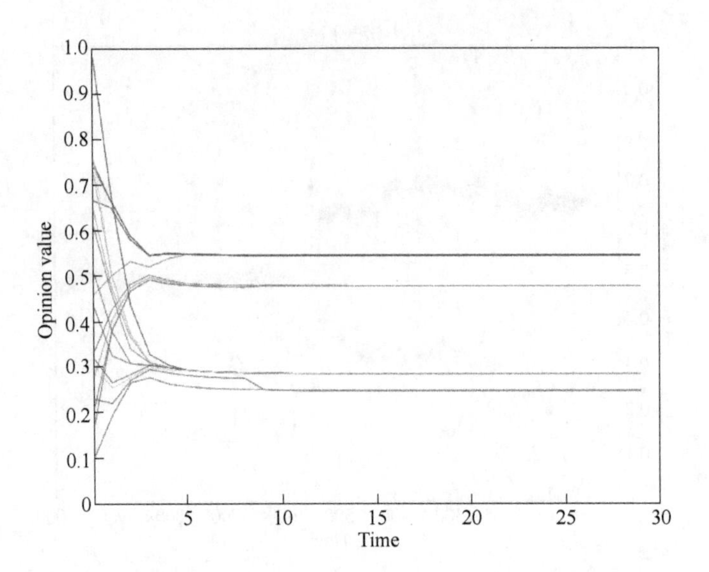

Fig. 8.3　Opinion evolution of system (8.1), (8.3), (8.9) of 20 agents without noise. 10 – 10 agents own the prejudice values J_1 and J_2 respectively

(The initial opinion values are randomly generated in [0,1], J_1 = 0.6, J_2 = 0.2, confidence threshold ε = 0.2, and attraction strength α = 0.4)

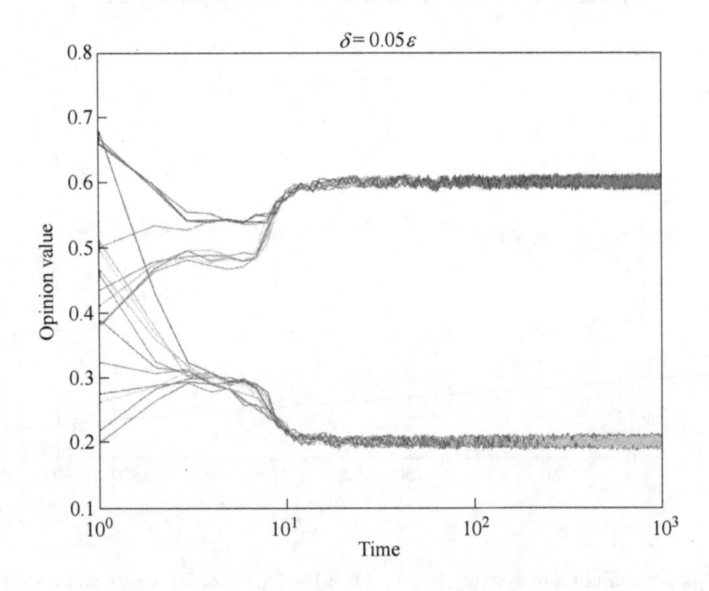

Fig. 8.4　Opinion evolution of system (8.1), (8.3), (8.9) of 20 agents with noise

(The initial conditions are the same as those in Fig. 8.3 and noise strength δ = 0.05ε)

8.4 Notes

In this chapter, we investigated how fragmentation phenomenon emerges with HK dynamics. It is shown that the original HK model and its variations with homogeneous prejudiced or stubborn agents fails to generate robust cleavage in the presence of tiny noise. We then revealed that the HK model with heterogeneous prejudiced agents can preserve a robust fragmentation in noisy environment. This implies a quite intuitive conclusion that it is the innate difference within group rather than only bounded confidence of individuals that causes the ubiquitous social cleavage. Moreover, we first find that noise could drive the prejudiced agents to their preferences and help the group form a bipartite cleavage. From a converse point of view, we can weaken the social cleavage by introducing some centrist strength to reduce the difference of heterogeneous prejudices in social group.

References

[1] Hegselmann R, Krause U. Opinion dynamics and bounded confidence models, analysis, and simulation[J]. Journal of Artificial Societies and Social Simulation, 2002, 5 (3): 1 ~ 33.

[2] Abelson R P. Mathematical models of the distribution of attitudes under controversy in Contributions to Mathematical Psychology[M]. N. Frederiksen and H. Gulliksen, Eds. New York: Holt Rinehart and Winston, 1964, 142 ~ 160.

[3] Friedkin N. The problem of social control and coordination of complex systems in sociology: a look at the community cleavage[J]. IEEE Control Systems, 2015, 35 (3): 40 ~ 51.

[4] Kurahashi-Nakamura T, Mäs M, Lorenza J. Robust clustering in generalized bounded confidence models[J]. J. Artificial Socie. Social Simul, 2016, 19 (4): 7.

[5] Deffuant G, Neau D, Amblard F, Weisbuch G. Mixing beliefs among interacting agents[J]. Advances in Complex Systems, 2000, 3 (01n04) :87 ~ 98.

[6] Dabarera R, Premaratne K, Murthi M, Sarkar D. Consensus in the presence of multiple opinion leaders effect of bounded Confidence[J]. IEEE Transactions on Signal and Information Processing over Networks, 2016, 2 (3): 336 ~ 349.

[7] Hegselmann R, Krause U. Truth and cognitive division of labour first steps towards a computer aided social epistemology[J]. Journal of Artificial Societies and Social Simulation, 2006, 9 (3): 1 ~ 28.

[8] Proskurnikov A, Tempo R, Cao M, Friedkin N. Opinion evolution in time-varying social influence networks with prejudiced agents[J]. IFAC Papers OnLine, 2017, 50 (1): 11896 ~ 11901.

[9] Su W, Yu Y. Free information flow benefits truth seeking[J]. Journal of Systems Science and Complexity, 2018, 31 (4), 964 ~ 974.

[10] Chen G. Small noise may diversify collective motion in Vicsek model[J]. IEEE Transactions on

Automatic Control, 2017, 62 (2): 636 ~651.

[11] Mäs M, Flache A, Helbing D. Individualization as driving force of clustering phenomena in humans[J]. PLoS Computational Biology, 2010, 6 (10), e1000959.

[12] Pineda M, Toral R, Hernndez-García E. Diffusing opinions in bounded confidence processes [J]. The European Physical Journal D, 2011, 62 (1): 109 ~117.

[13] Guo J, Mu B, Wang L, Yin G, Xu L. Decision-Based System Identification and Adaptive Resource Allocation[J]. IEEE Transactions on Automatic Control, 2017, 62 (5): 2166 ~2179.

[14] Li S, Batra R, Brown D, Chang H, Ranganathan N, Hoberman C, Rus D, Lipson H. Particle robotics based on statistical mechanics of loosely coupled components[J]. Nature, 2019, 567, 361 ~365.

[15] Su W, Chen G, Yu Y. Finite-Time Elimination of Disagreement of Opinion Dynamics via Covert Noise[J]. IET Control Theory and Applications, 2018, 12 (4): 563 ~570.

[16] Krause U. A discrete nonlinear and non-autonomous model of consensus formation[M]. Communications in Difference Equations, Amsterdam: Gordon and Breach Publisher, 2000, 227 ~238.

[17] Chazelle B, Wang C. Inertial Hegselmann-Krause systems[J]. IEEE Transactions on Automatic Control, 2015, 62 (8): 3905 ~3913.

[18] Von Foerster H. On self-organizing systems and their environments, Self-organizing systems[J]. Pergamon Press, London, 1960, 31 ~50.

Appendix A Proofs

A.1 Proof of Lemma 5.2

Proof We consider the protocol (5.10) first. Without loss of generality we assume $\alpha' \in \left(0, \frac{\eta}{2}\right)$. The main idea of this proof is: For each agent i, if its neighbors' average opinion $\tilde{x}_i(t)$ is larger than an upper bound, we set $u_i(t)$ to be a negative input; if $x_i(t)$ is less than a lower bound, we set $u_i(t)$ to be a positive input. Otherwise, we select a control input such that $x_i(t+1)$ will be in the interval $[z'-\alpha', z'+\alpha']$. With this idea, for $t \geqslant 0$ and $1 \leqslant i \leqslant n$ we choose $\delta_i(t) = \alpha'$ and

$$u_i(t) = \begin{cases} -\eta + \alpha' & \text{if } x_i(t) > z' + \eta - \alpha' \\ z' - x_i(t) & \text{if } x_i(t) \in [z' - \eta + \alpha',\ z' + \eta - \alpha'] \\ \eta - \alpha' & \text{if } x_i(t) < z' - \eta + \alpha' \end{cases}$$

It can be computed that

$$u_i(t) \in [-\eta + \delta_i(t),\ \eta - \delta_i(t)],\ \forall i \in V,\ t \geqslant 0$$

which meets its requirement in Definition 5.2. Define

$$x_{\max}(t) \triangleq \max_{1 \leqslant i \leqslant n} x_i(t) \quad \text{and} \quad x_{\min}(t) \triangleq \min_{1 \leqslant i \leqslant n} x_i(t)$$

For any $i \in V$, if $x_i(t) > z' + \eta - \alpha'$ which indicates $x_{\max}(t) > z' + \eta - \alpha'$, by (A.1) we have

$$\begin{aligned} x_i(t+1) &= x_i(t) + u_i(t) + b_i(t) = x_i(t) - \eta + \alpha' + b_i(t) \\ &\in (z' + b_i(t),\ x_{\max}(t) - \eta + \alpha' + b_i(t)] \end{aligned} \tag{A.1}$$

and

$$(z' + b_i(t),\ x_{\max}(t) - \eta + \alpha' + b_i(t)] \subseteq (z' - \alpha',\ x_{\max}(t) - \eta + 2\alpha'] \tag{A.2}$$

If $x_i(t) < z' - \eta + \alpha'$ which indicates $x_{\min}(t) < z' - \eta + \alpha'$, similar to (A.1) and (A.2) we have

$$x_i(t+1) \in [x_{\min}(t)+\eta-2\alpha',\ z'+\alpha') \tag{A.3}$$

Otherwise if $x_i(t) \in [z'-\eta+\alpha',\ z'+\eta-\alpha']$, by (A.1) we have

$$x_i(t+1) = x_i(t)+z'-x_i(t)+b_i(t) \in [z'-\alpha',\ z'+\alpha'] \tag{A.4}$$

From (A.1) – (A.4) together we can get

$$x_{\max}(t+1) \leqslant \max\{x_{\max}(t)-\eta+2\alpha',\ z'+\alpha'\}$$

and

$$x_{\min}(t+1) \geqslant \min\{x_{\min}(t)+\eta-2\alpha',\ z'-\alpha'\}$$

so there must exist a finite time t_1 such that $x_i(t_1) \in [z'-\alpha',\ z'+\alpha']$ for all $i \in V$.

For protocol (5.10), this result can be obtained by a similar method to the above.

□

A.2 Proof of Lemma 5.3

Proof Let K be a constant satisfying

$$K \geqslant \max\left\{\frac{4(r_{\min}+\eta)}{2\eta-r_{\min}},\ \frac{2n\eta}{n\eta-r_{\max}}\right\}$$

By Lemma 5.2 the set $S_{\frac{K+2}{2K}r_{\min},\ \frac{r_{\min}}{K}}$ is finite-time robustly reachable from $[0,1]^n$ under protocol (5.10). We record the stop time of the set $S_{\frac{K+2}{2K}r_{\min},\ \frac{r_{\min}}{K}}$ being reached as t_1, which means

$$x_i(t_1) \in \left[\frac{r_{\min}}{2},\ \frac{r_{\min}}{2}+\frac{2r_{\min}}{K}\right],\ \forall i \in V$$

By this and (5.9) we have

$$x_1(t_1) = x_2(t_1) = \cdots = x_n(t_1) \in \left[\frac{r_{\min}}{2},\ \frac{r_{\min}}{2}+\frac{2r_{\min}}{K}\right] \tag{A.5}$$

Without loss of generality we assume agent 1 has the smallest interaction radius, i. e., $r_1 = r_{\min}$. For any $t \geqslant t_1$ and $i \in V$ we choose $\delta_i(t) = \frac{\eta}{K}$, and

$$u_i(t) = \begin{cases} \eta - \dfrac{\eta}{K} & \text{if } i = 1 \\ -\eta + \dfrac{\eta}{K} & \text{otherwise} \end{cases} \tag{A.6}$$

Because $x_i(t_1+1) = \prod_{[0,1]}(x_i(t_1)+u_i(t_1)+b_i(t_1))$, by this and (A.5) we can get

$$
\begin{aligned}
x_1(t_1+1) &\geqslant \min\left\{x_1(t_1)+\eta-\frac{2\eta}{K},\ 1\right\} \\
&\geqslant \min\left\{\frac{r_{\min}}{2}+\eta-\frac{2\eta}{K},\ 1\right\} > r_{\min}
\end{aligned}
\tag{A.7}
$$

and

$$
\begin{aligned}
x_i(t_1+1) &\leqslant \max\left\{x_i(t_1)-\eta+\frac{2\eta}{K},\ 0\right\} \\
&\leqslant \max\left\{\frac{r_{\min}}{2}+\frac{2r_{\min}}{K}-\eta+\frac{2\eta}{K},\ 0\right\} \\
&= 0,\ i=2,\cdots,n
\end{aligned}
\tag{A.8}
$$

so

$$
x_1(t_1+1)-x_i(t_1+1) > r_{\min} = r_1,\ i=2,\cdots,n
$$

Next we compute $x_i(t_1+2)$. Because agent 1 cannot receive information from the others at time t_1+1, with (A.7) we get

$$
\begin{aligned}
x_1(t_1+2) &= x_1(t_1+1)+u_i(t_1+1)+b_i(t_1+1) \\
&\geqslant \min\left\{x_1(t_1)+2\left(\eta-\frac{2\eta}{K}\right),\ 1\right\}
\end{aligned}
\tag{A.9}
$$

Also, for $2\leqslant i\leqslant n$, by (A.8) we have

$$
x_i(t_1+1) = \frac{x_1(t_1+1)}{n} I_{\{x_1(t_1+1)\leqslant r_i\}} \leqslant \frac{r_i}{n} \leqslant \frac{r_{\max}}{n}
$$

which is followed by

$$
x_i(t_1+2) \leqslant \max\left\{\frac{r_{\max}}{n}-\eta+\frac{2\eta}{K},\ 0\right\} = 0
\tag{A.10}
$$

Repeating the process of (A.9) ~ (A.10) we get that there exists a finite time $t_2 > t_1$ such that $x_1(t_2)=1$, and $x_i(t_2)=0$ for $2\leqslant i\leqslant n$. □

A.3 Proof of Lemma 5.4

Proof We consider protocol (5.10) first. Without loss of generality we assume that ε

is arbitrarily small (though positive). By Lemma 5.2 the set $S_{\frac{1}{2},\frac{\varepsilon}{2}}$ is finite-time robustly reachable from $[0,1]^n$ under protocol (5.10). We record the stop time of the set $S_{\frac{1}{2},\frac{\varepsilon}{2}}$ being reached as t_1, which implies

$$x_1(t_1)=x_2(t_1)\in\left[\frac{1}{2}-\frac{\varepsilon}{2},\ \frac{1}{2}+\frac{\varepsilon}{2}\right] \tag{A.11}$$

We choose $\delta_i(t_1)=\frac{\varepsilon}{4}$ for any $i\in V$, and $u_1(t_1)=\eta-\frac{\varepsilon}{4}$, and $u_2(t_1)=-\eta+\frac{\varepsilon}{4}$. With this and (A.11) we have

$$\begin{aligned}x_1(t_1+1)&\geqslant\min\{x_1(t_1)+u_1(t_1)-\delta_1(t_1),1\}\\&\geqslant\min\left\{x_1(t_1)+\eta-\frac{\varepsilon}{2},1\right\}\end{aligned}$$

and

$$\begin{aligned}x_2(t_1+1)&\leqslant\max\{x_2(t_1)+u_2(t_1)+\delta_2(t_1),0\}\\&\leqslant\max\left\{x_2(t_1)-\eta+\frac{\varepsilon}{2},0\right\}\end{aligned}$$

By these and (A.11) we get $x_1(t_1+1)-x_2(t_1+1)\geqslant c_\varepsilon$.

For protocol (5.10), our result can be obtained by a similar method to the above. □

A.4 Proof of Lemma 5.5

Proof Without loss of generality we assume $r_1<2\eta$ and $\omega_1=\underline{\omega}_\eta$. For any constant K satisfying

$$K\geqslant\max\left\{\frac{4(r_1+\eta)}{2\eta-r_1},\ \frac{4\eta n}{(n-1)\omega_1\varepsilon},\ \frac{2\eta n}{\varepsilon}\right\} \tag{A.12}$$

with a similar method to the proof of Lemma 5.3 we can find a finite time t_1 such that

$$x_1(t_1)=x_2(t_1)=\cdots=x_n(t_1)\in\left[\frac{r_1}{2},\ \frac{r_1}{2}+\frac{2r_1}{K}\right] \tag{A.13}$$

For any $t\geqslant t_1$ and $i\in V$ we choose $\delta_i(t)=\frac{\eta}{K}$, and $u_i(t)$ as same as (A.6). Because $x_i(t_1+1)=\prod_{[0,1]}(x_i(t_1)+u_i(t_1)+b_i(t_1))$, similar to (A.7) and (A.8) we have

$$x_1(t_1+1)\geqslant\min\left\{x_1(t_1)+\eta-\frac{2\eta}{K},1\right\}>r_1$$

and

$$x_i(t_1+1)\leqslant\max\left\{x_i(t_1)-\eta+\frac{2\eta}{K},\ 0\right\}=0,\ i=2,\cdots,n$$

From these we get

$$\begin{aligned}x_1(t_1+2)&\geqslant\min\left\{x_1(t_1+1)+\eta-\frac{2\eta}{K},\ 1\right\}\\&=\min\left\{\omega_1x_{\text{ave}}(t_1+1)+(1-\omega_1)x_1(t_1+1)+\eta-\frac{2\eta}{K},\ 1\right\}\\&=\min\left\{\left(1-\omega_1+\frac{\omega_1}{n}\right)x_1(t_1+1)+\eta-\frac{2\eta}{K},\ 1\right\}\end{aligned}$$

and for $2\leqslant i\leqslant n$ we have

$$\begin{aligned}x_i(t_1+2)&\leqslant\max\left\{x_i(t_1+1)-\eta+\frac{2\eta}{K},\ 0\right\}\\&\leqslant\max\left\{\frac{x_1(t_1+1)}{n}-\eta+\frac{2\eta}{K},\ 0\right\}\end{aligned}\tag{A.14}$$

Also, for any $y<\min\left\{\frac{n\eta}{(n-1)\omega_1}-\varepsilon,\ n\eta-\varepsilon\right\}$, by (A.12) we have

$$\begin{aligned}\left(1-\omega_1+\frac{\omega_1}{n}\right)y+\eta-\frac{2\eta}{K}&>y-\frac{(n-1)\omega_1}{n}\left(\frac{n\eta}{(n-1)\omega_1}-\varepsilon\right)+\eta-\frac{2\eta}{K}\\&\geqslant y+\frac{(n-1)\omega_1\varepsilon}{2n}\end{aligned}\tag{A.15}$$

and

$$\frac{y}{n}-\eta+\frac{2\eta}{K}<\frac{n\eta-\varepsilon}{n}-\eta+\frac{2\eta}{K}\leqslant 0\tag{A.16}$$

Taking (A.15) into (A.14), and (A.16) into (A.14), and repeatedly computing $x_i(t_1+3)$, $x_i(t_1+4)$, $\cdots$, there must exist a finite time $t_2>t_1$ such that

$$x_1(t_2)-x_i(t_2)\geqslant\min\left\{\frac{n\eta}{(n-1)\omega}-\varepsilon,\ n\eta-\varepsilon,\ 1\right\}$$

for all $i\in\{2,\cdots,n\}$. □

A.5 Proof of Lemma 5.8

Proof Without loss of generality we assume $\omega_1+\omega_2=\min_{i\neq j}(\omega_i+\omega_j)$, and ε is arbitrarily small (A. though positive). Let

$$x^*: = \prod_{[0,1]}\left(\frac{1}{2} + \frac{(\omega_1 - \omega_2)(n-1)\eta}{2n}\right)$$

By Lemma 5.7 the set $S_{x^*,\frac{\varepsilon}{2}}$ is finite-time robustly reachable from $[0,1]^n$ under protocol (5.14). We record the stop time of the set $S_{x^*,\frac{\varepsilon}{2}}$ being reached as t_1, which implies

$$x_1(t_1) = \cdots = x_n(t_1) \in \left[x^* - \frac{\varepsilon}{2},\ x^* + \frac{\varepsilon}{2}\right] \tag{A.17}$$

For $i = 1,2$, let $\varepsilon_i \triangleq \frac{n\varepsilon}{(1-\omega_i)(n-1)}$, and choose $\delta_i(t_1) = \frac{\varepsilon_i}{4}$. Also, choose $u_{j1}(t_1) = \eta - \frac{\varepsilon_1}{4}$ for any $j \in N_1(t_1)\setminus\{1\}$ and $u_{j2}(t_1) = -\eta + \frac{\varepsilon_2}{4}$ for any $j \in N_2(t_1)\setminus\{2\}$. Combining this with (5.14) and (A.17) we have

$$\begin{aligned} x_1(t_1+1) &= \prod_{[0,1]}\left(x_1(t) + \frac{1-\omega_1}{n}\sum_{j\neq 1}(u_{j1}(t) + b_{j1}(t))\right) \\ &\geqslant \min\left\{x_1(t_1) + \frac{(1-\omega_1)(n-1)}{n}\left(\eta - \frac{\varepsilon_1}{2}\right),\ 1\right\} \\ &\geqslant \min\left\{x^* + \frac{(1-\omega_1)(n-1)\eta}{n} - \varepsilon,\ 1\right\} \end{aligned}$$

and

$$\begin{aligned} x_2(t_1+1) &= \prod_{[0,1]}\left(x_2(t) + \frac{1-\omega_2}{n}\sum_{j\neq 2}(u_{j2}(t) + b_{j2}(t))\right) \\ &\leqslant \max\left\{x_2(t_1) + \frac{(1-\omega_2)(n-1)}{n}\left(-\eta + \frac{\varepsilon_2}{2}\right),\ 0\right\} \\ &\leqslant \max\left\{x^* - \frac{(1-\omega_2)(n-1)\eta}{n} + \varepsilon,\ 0\right\} \end{aligned}$$

Thus we get

$$\begin{aligned} x_1(t_1+1) - x_2(t_1+1) &\geqslant \min\left\{\frac{(2-\omega_1-\omega_2)(n-1)\eta}{n} - 2\varepsilon,\ 1\right\} \\ &= \min\{a_{12} - 2\varepsilon,\ 1\} = c_\varepsilon \end{aligned} \tag{A.18}$$

□

A.6 Proof of Lemma 5.9

Proof We first show that $E_{c_{12}-\varepsilon}$ is finite-time robustly reachable from $[0,1]^n$. By a similar method to the proof of Lemma 5.3, for any real number $x^* \in [0,1]$, we can find a finite time t_1 such that

$$x_1(t_1)=x_2(t_1)=\cdots=x_n(t_1)\in\left[x^*-\frac{\eta}{K},\ x^*+\frac{\eta}{K}\right] \tag{A.19}$$

where $K>0$ is a constant large enough.

At the time t_1, we choose $\delta_1(t_1)=\delta_2(t_1)=\frac{\eta}{K}$, and $\delta_3(t_1)=\cdots=\delta_n(t_1)=\frac{\eta}{MK}$ with M be a large integer. Also, for $i\in V$ and $j\in N_i(t_1)\backslash\{i\}$ we choose

$$u_{ji}(t_1)=\begin{cases}\eta-\dfrac{2\eta}{K} & \text{if } i=1\\[2mm] -\eta+\dfrac{2\eta}{K} & \text{if } i=2\\[2mm] 0 & \text{if } 3\leqslant i\leqslant n\end{cases} \tag{A.20}$$

By (5.14), (A.19) and (A.20) we have

$$\begin{cases}x_1(t_1+1)\geqslant\min\left\{x_1(t_1)+\dfrac{(n-1)(1-\omega_1)}{n}\left(\eta-\dfrac{3\eta}{K}\right),\ 1\right\}\\ x_1(t_1+1)\leqslant\min\left\{x_1(t_1)+\dfrac{(n-1)(1-\omega_1)}{n}\left(\eta-\dfrac{\eta}{K}\right),\ 1\right\}\\ x_2(t_1+1)\geqslant\max\left\{x_1(t_1)-\dfrac{(n-1)(1-\omega_2)}{n}\left(\eta-\dfrac{\eta}{K}\right),\ 0\right\}\\ x_2(t_1+1)\leqslant\max\left\{x_1(t_1)-\dfrac{(n-1)(1-\omega_2)}{n}\left(\eta-\dfrac{3\eta}{K}\right),\ 0\right\}\\ x_i(t_1+1)\geqslant\min\left\{x_1(t_1)-\dfrac{(n-1)(1-\omega_i)\eta}{nMK},\ 0\right\},\ 3\leqslant i\leqslant n\\ x_i(t_1+1)\leqslant\max\left\{x_1(t_1)+\dfrac{(n-1)(1-\omega_i)\eta}{nMK},\ 1\right\},\ 3\leqslant i\leqslant n\end{cases} \tag{A.21}$$

If $a_{12}=\dfrac{(n-1)(2-\omega_1-\omega_2)\eta}{n}\geqslant 1$, similar to (A.18) we can choose suitable x^* and large K such that

$$x_1(t_1+1)-x_2(t_1+1)\geqslant\min\{a_{12}-\varepsilon,\ 1\}\geqslant 1-\varepsilon\geqslant c_{12}-\varepsilon$$

which indicates that $E_{c_{12}-\varepsilon}$ is robustly reached at time t_1+1. Thus, we just need to consider the case of $a_{12}<1$. In this case we can choose suitable x^* and large K such that

$$x_1(t_1)+a_1<1 \quad \text{and} \quad x_1(t_1)-a_2>0 \tag{A.22}$$

Here we recall that $a_i=\dfrac{(n-1)(1-\omega_i)\eta}{n}$. Also, we can get

$$x_{\text{ave}}(t_1+1)\approx x_1(t_1)+\frac{(n-1)(\omega_2-\omega_1)\eta}{n^2} \tag{A.23}$$

where $A\approx B$ indicates that $\lim\limits_{K\to\infty}(A-B)=0$ in this proof.

At the time t_1+1, we choose $\delta_1(t_1+1)=\delta_2(t_1+1)=\dfrac{\eta}{K}$, and $u_{j1}(t_1+1)=\eta-\dfrac{\eta}{K}$ for $j\in N_1(t_1+1)\setminus\{1\}$, while $u_{j2}(t_1+1)=-\eta+\dfrac{\eta}{K}$ for $j\in N_2(t_1+1)\setminus\{2\}$.

If $r_1<a_1$, by (A.21) we can get $N_1(t_1+1)=\{1\}$ for large K, so by (5.14)

$$\begin{aligned} x_1(t_1+2) &= \omega_1 x_{\text{ave}}(t_1+1)+(1-\omega_1)x_1(t_1+1) \\ &\approx x_1(t_1)+\frac{(n-1)\eta}{n}\Big[(1-\omega_1)^2+\frac{\omega_1(\omega_2-\omega_1)}{n}\Big] \end{aligned} \tag{A.24}$$

If $r_1\in[a_1,a_{12})$, by (A.21) we can get $N_1(t_1+1)=\{1,3,\cdots,n\}$ for large K and M, so by (5.14)

$$\begin{aligned} x_1(t_1+2) &\approx \min\Big\{\omega_1 x_{\text{ave}}(t_1+1)+\frac{1-\omega_1}{n-1}[(n-2)x_1(t_1)+x_1(t_1+1)+(n-2)\eta],\ 1\Big\} \\ &\approx \min\Big\{x_1(t_1)+\frac{\omega_1(n-1)(\omega_2-\omega_1)\eta}{n^2}+(1-\omega_1)\Big[\frac{(1-\omega_1)\eta}{n}+\frac{(n-2)\eta}{n-1}\Big],\ 1\Big\} \end{aligned} \tag{A.25}$$

If $r_1\geqslant a_{12}$, by (A.21) we can get $N_1(t_1+1)=\{1,2,\cdots,n\}$ for large K and M, so by (5.14) and (A.23)

$$\begin{aligned} x_1(t_1+2) &\approx \min\Big\{\omega_1 x_{\text{ave}}(t_1+1)+(1-\omega_1)\Big[x_{\text{ave}}(t_1+1)+\frac{(n-1)\eta}{n}\Big],\ 1\Big\} \\ &\approx \min\Big\{x_1(t_1)+\frac{(n-1)\eta}{n}\Big(1-\omega_1+\frac{\omega_2-\omega_1}{n}\Big),\ 1\Big\} \end{aligned} \tag{A.26}$$

From (A.24), (A.25), (A.26) and (5.9) we get

$$x_1(t_1+2) \approx \prod_{[0,1]}(x_1(t_1)+h_{12}) \tag{A.27}$$

By a similar discussion we can get

$$x_2(t_1+2) \approx \prod_{[0,1]}(x_1(t_1)-h_{21}).\zeta_{ji}(t) \in [-\eta,\eta] \tag{A.28}$$

Let $A \lesssim B$ denote $\lim_{K\to\infty}(A-B) \leqslant 0$. If $a_1 \leqslant h_{12}$ and $a_2 \leqslant h_{21}$, which means $x_1(t_1+1) \lesssim x_1(t_1+2)$ and $x_2(t_1+2) \lesssim x_2(t_1+1)$, then we choose x^* to be h_{21} or $1-h_{12}$ and get $x_1(t_1+2) - x_2(t_1+2) \approx \min\{h_{12}+h_{21}, 1\}$, which indicates $E_{c12-\varepsilon}$ is robustly reached at time t_1+2.

If $a_1 > h_{12}$ and $a_2 \leqslant h_{12}$, we choose $x^* = 1-a_1$ which means $x_1(t_1+1) \approx 1$, and so $x_1(t_1+2) - x_2(t_1+2) \approx \min\{h_{12}+h_{21}, 1-(a_1-h_{12})\}$, which indicates $E_{c12-\varepsilon}$ is also robustly reached at time t_1+2.

If $a_1 \leqslant h_{12}$ and $a_2 > h_{21}$, we choose $x^* = a_2$ and get $x_1(t_1+2) - x_2(t_1+2) \approx \min\{h_{12}+h_{21}, 1-(a_2-h_{21})\}$, so $E_{c12-\varepsilon}$ is also robustly reached at time t_1+2.

If $a_1 > h_{12}$ and $a_2 > h_{12}$, we have $h_{12}+h_{21} < x_1(t_1+1) - x_2(t_1+1)$, so $E_{c12-\varepsilon}$ is robustly reached at time t_1+1.

Given the discussion above, $E_{c12-\varepsilon}$ is finite-time robustly reachable from $[0,1]^n$.

For any $i \neq j$, by a similar method we have $E_{cij-\varepsilon}$ is finite-time robustly reachable from $[0,1]^n$, so E_{c_ε} is finite-time robustly reachable from $[0,1]^n$. □

A.7 Proof of Lemma 5.10

Proof Set $x_{\max}(t)$ and $x_{\min}(t)$ to be the maximal and minimum opinions at time t respectively. Let $K>0$ be a large constant and set $\delta_i(t) = \frac{\eta}{K}$ for $i \in V$ and $t \geqslant 0$. Next we try to find a control algorithm such that

$$x_{\max}(t_1) - x_{\min}(t_1) < \max\left\{x_{\max}(0) - x_{\min}(0) - \frac{\eta}{2} + \frac{2\eta}{K},\ x_{\max}(0) - x_{\min}(0) - r_1,\ \frac{2\eta}{K}\right\} \tag{A.29}$$

where t_1 is a finite time. Assume $x_{\max}(0) - x_{\min}(0) > \frac{2\eta}{K}$. Because for any initial opinions, there must exist two agents whose distance is not bigger than $\frac{1}{n-1}$, which means they are not isolated, we prove (A.29) for the following three cases respectively:

Case Ⅰ: If the agent with the minimum opinion is not isolated, for all $i \in V$, $j \in N_i(0) \setminus \{i\}$ we choose

$$u_{ji}(0) = \min\left\{\eta, \frac{[x_{\max}(0) - x_i(0)]\,|N_i(0)|}{|N_i(0)| - 1}\right\} - \frac{\eta}{K} \quad (A.30)$$

By (5.14) and the fact that $x_{\min}(t) \leqslant x_i(t) \leqslant x_{\max}(t)$, we can get for any $i \in V$ satisfying $|N_i(0)| \geqslant 2$,

$$\begin{aligned} x_i(1) &\geqslant \min\left\{x_{\min}(0) + \frac{|N_i(0)| - 1}{|N_i(0)|}\left[\eta - \frac{2\eta}{K}\right], x_{\max}(0) - \frac{|N_i(0)| - 1}{|N_i(0)|}\frac{2\eta}{K}\right\} \\ &> \min\left\{x_{\min}(0) + \frac{\eta}{2} - \frac{2\eta}{K}, x_{\max}(0) - \frac{2\eta}{K}\right\} \end{aligned}$$

Also, $x_i(1) \leqslant x_{\max}(0)$ for all $i \in V$, and if $|N_i(0)| = 1$ then $x_i(0) > x_{\min}(0) + r_1$, so (A.29) holds when $t_1 = 1$.

Case Ⅱ: If the agent with the maximal opinion is not isolated, for all $i \in V$, $j \in N_i(0) \setminus \{i\}$ we choose

$$u_{ji}(0) = \max\left\{-\eta, \frac{[x_{\min}(0) - x_i(0)]\,|N_i(0)|}{|N_i(0)| - 1}\right\} + \frac{\eta}{K} \quad (A.31)$$

Similar to Case Ⅰ we get (A.29) holds when $t_1 = 1$.

Case Ⅲ: If the agents with the minimum and maximal opinions are all isolated, for all $i \in V$, $j \in N_i(0) \setminus \{i\}$ we choose

$$u_{ji}(t) = \min\left\{\eta, \frac{[x_{\max}(t) - x_i(t)]\,|N_i(t)|}{|N_i(t)| - 1}\right\} - \frac{\eta}{K} \quad (A.32)$$

until the agent with the maximal opinion is not isolated. Let $y(t)$ be the minimal value of the non-isolated agents' opinions. Under (A.32) we have

$$\min\left\{y(t) + \frac{\eta}{2} - \frac{2\eta}{K}, x_{\max}(t) - \frac{2\eta}{K}\right\} \leqslant y(t+1) \leqslant x_{\max}(t), \quad \forall i \in V$$

so there exists a finite time t_0 such that the agent with the maximal opinion is not isolated at time t'. With the same method as Case Ⅱ we can get (A.29) holds when $t_1 = t_0 + 1$.

Repeatedly using (A.29) we get that there exists a finite time t' such that $x_{\max}(t') - x_{\min}(t') \leqslant \frac{2\eta}{K}$. Finally we design a control algorithm which moves $(x_{\min}(t') + x_{\max}(t'))/2$ to z^* while $x_{\max}(t) - x_{\min}(t)$ keeps not bigger than $\frac{2\eta}{K}$. Set $x'(t) \triangleq \frac{x_{\min}(t) + x_{\max}(t)}{2}$.

For any $t \geqslant t'$, $i \in V$, and $j \in N_i(t) \setminus \{i\}$ we choose

$$u_{ji}(t) = \begin{cases} \dfrac{[x'(t) - x_i(t)]n}{n-1} - \eta + \dfrac{2\eta}{K}, & \text{if } x'(t) > z^* + \dfrac{n-1}{n}\left(\eta - \dfrac{2\eta}{K}\right) \\ \dfrac{[z^* - x_i(t)]n}{n-1}, & \text{if } x_i(t) \in \left[z^* - \dfrac{n-1}{n}\left(\eta - \dfrac{2\eta}{K}\right), z^* + \dfrac{n-1}{n}\left(\eta - \dfrac{2\eta}{K}\right)\right] \\ \dfrac{[x'(t) - x_i(t)]n}{n-1} + \eta - \dfrac{2\eta}{K}, & \text{if } x'(t) < z^* - \dfrac{n-1}{n}\left(\eta - \dfrac{2\eta}{K}\right) \end{cases}$$

By this and (5. 14) we have $x'(t+1) \leqslant x'(t) - \frac{n-1}{n}\left(\eta - \frac{3\eta}{K}\right)$ if $x'(t) > z^* + \frac{n-1}{n}\left(\eta - \frac{2\eta}{K}\right)$, and $x'(t+1) \geqslant x'(t) + \frac{n-1}{n}\left(\eta - \frac{3\eta}{K}\right)$ if $x'(t) < z^* - \frac{n-1}{n}\left(\eta - \frac{2\eta}{K}\right)$. Then, there exists a finite time t^* such that

$$\begin{aligned} x_i(t^*) &= x_i(t^* - 1) + \frac{1}{n}\sum_{j \neq i}\left(\frac{[z^* - x_i(t^* - 1)]n}{n-1} + b_{ji}(t^* - 1)\right) \\ &= z^* + \frac{1}{n}\sum_{j \neq i} b_{ji}(t^* - 1) \end{aligned}$$

which indicates $|x_i(t^*) - z^*| < \frac{\eta}{K}$ for all $i \in V$ by $|b_{ji}(t)| \leqslant \frac{\eta}{K}$. Thus, S_{z^*, α^*} is finite-time robustly reachable from $[0,1]^n$ if we let $K \geqslant \lceil \eta/\alpha^* \rceil$. □

A. 8 Proof of Lemma 5. 11

Proof Let $K > 0$ be a large constant. By Lemma 5. 10, the set hhh $S_{\frac{1}{2}, \frac{r_1}{K}}$ is finite-time robustly reachable from $[0,1]^n$ under protocol (5. 14). We record the stop time of the set $S_{\frac{1}{2}, \frac{r_1}{K}}$ being reached as t_1, which means

$$x_i(t_1) \in \left[\frac{1}{2} - \frac{r_1}{K}, \frac{1}{2} + \frac{r_1}{K}\right], \quad \forall i \in V \tag{A.33}$$

Then $|N_i(t_1)| = n$ and $x_i(t_1) = \sum_{j=1}^{n} \frac{x_j(t_1)}{n}$ for all $i \in V$. Choose $\delta_i(t_1) = \frac{\eta}{K}$, and

$$u_{ji}(t_1) = \begin{cases} -\eta + \dfrac{\eta}{K} & \text{if } i = 1, j \in N_1(t) \setminus \{1\} \\ \eta - \dfrac{\eta}{K} & \text{if } 2 \leqslant i \leqslant n, j \in N_i(t) \setminus \{i\} \end{cases} \tag{A.34}$$

then by (5. 14) we have

$$x_1(t_1+1) \in \left[x_1(t_1) - \frac{n-1}{n}\eta,\ x_1(t_1) - \frac{n-1}{n}\left(\eta - \frac{\eta}{K}\right)\right]$$

and

$$x_i(t_1+1) \in \left[x_i(t_1) + \frac{n-1}{n}\left(\eta - \frac{\eta}{K}\right),\ x_i(t_1) + \frac{n-1}{n}\eta\right]$$

for all $i \in \{2,\cdots,n\}$. Thus, E'_ε is robustly reached at time t_1+1. □

A.9 Proof of Lemma 5.12

Proof Similar to (A.33) there exists a time t_1 such that

$$x_i(t_1) \in \left[\frac{r_1}{2} - \frac{r_1}{K},\ \frac{r_1}{2} + \frac{r_1}{K}\right],\ \forall i \in V$$

where $K>0$ is a large constant. For all $t \geqslant t_1$ and $i \in V$ we choose $\delta_i(t_1) = \frac{\eta}{K}$, and

$$u_{ji}(t) = \begin{cases} -\eta + \dfrac{\eta}{K} & \text{if } i=1,\ j \in N_1(t) \setminus \{1\} \\ \eta - \dfrac{\eta}{K} & \text{if } 2 \leqslant i \leqslant n,\ j \in N_i(t) \setminus \{i\} \end{cases} \tag{A.35}$$

then

$$x_i(t_1) = \sum_{j=1}^{n} \frac{x_j(t_1)}{n} \in \left[\frac{r_1}{2} - \frac{r_1}{K},\ \frac{r_1}{2} + \frac{r_1}{K}\right],\ \forall i \in V$$

Also, by (5.14) and the fact that $\eta > \frac{nr_1}{2(n-1)}$ we get

$$\begin{aligned} x_1(t_1+1) &\leqslant \prod_{[0,1]}\left(x_1(t_1) - \frac{n-1}{n}\left(\eta - \frac{\eta}{K}\right)\right) \\ &\leqslant \prod_{[0,1]}\left(\frac{r_1}{2} + \frac{r_1}{K} - \frac{n-1}{n}\left(\eta - \frac{\eta}{K}\right)\right) = 0 \end{aligned}$$

and for all $i \in \{2,\cdots,n\}$

$$x_i(t_1+1) \in \left[x_i(t_1) + \frac{n-1}{n}\left(\eta - \frac{\eta}{K}\right),\ x_i(t_1) + \frac{n-1}{n}\eta\right]$$

and

$$\left[x_i(t_1)+\frac{n-1}{n}\left(\eta-\frac{\eta}{K}\right),\ x_i(t_1)+\frac{n-1}{n}\eta\right]\subset\left[\frac{r_1}{2}-\frac{r_1}{K}+\frac{n-1}{n}\left(\eta-\frac{\eta}{K}\right),\ \frac{r_1}{2}+\frac{r_1}{K}+\frac{n-1}{n}\eta\right]$$

These yield that $x_i(t_1+1)-x_1(t_1+1)>r_1$ for all $i=2,\cdots,n$, and $x_2(t_1+1),\cdots,x_n(t_1+1)$ are neighbors to each other. Using (A.35) repeatedly, there exists a finite time $t_2>t_1$ such that $x_1(t_2)=0$ and $x_2(t_2)=\cdots=x_n(t_2)=1$.